International Nuclear Science Technology and Engineering Conference 2022 (iNuSTEC 2022)

IOP Conference Series: Materials Science and Engineering
Volume 1285

Bangi, Malaysia
25-27 October 2022

Editors:

Faridah Mohamad Idris
Julie Andrianny Murshidi
Faizal Mohamed
Irman Abdul Rahman
Azman Jalar
Norsyahidah Mohd Hidzir
Syazwani Mohd Fadzil
Nurul Nafiqah Itam Sulaiman
Mohd Idzat Idris
Aznan Fazli Ismail
Imran Yusuff
Khoo Kok Siong
Hoh Siew Yan
Hassan Mohamed
Abdul Aziz Mohamed

ISBN: 978-1-7138-7725-7
ISSN: 1757-8981

Printed from e-media with permission by:

Curran Associates, Inc.
57 Morehouse Lane
Red Hook, NY 12571

Some format issues inherent in the e-media version may also appear in this print version.

Institute of Physics
Dirac House, Temple Back
Bristol BS1 6BE UK

Phone: 44 1 17 929 7481
Fax: 44 1 17 920 0979

techtracking@iop.org

Additional copies of this publication are available from:

Curran Associates, Inc.
57 Morehouse Lane
Red Hook, NY 12571 USA
Phone: 845-758-0400
Fax: 845-758-2633
Email: curran@proceedings.com
Web: www.proceedings.com

TABLE OF CONTENTS

Preface

Peer Review Statement

Pragmatic Modelling of Axial Flow-Induced Vibration (FIV) for Nuclear Fuel Rods 1
Anas Muhamad Pauzi, Hector Iacovides, Andrea Cioncolini

Improving Source Convergence Diagnostics Sensitivity Using the Complex Components of Fourier Fundamental Mode 7
K F Chin, M R Omar, J A Karim

Nuclear Photonics: Laser-Driven Nuclear Physics 15
J. F. Ong, Meng-Hock Koh, I. H. Hashim

Study the Closed Orbit Correction of a Compact Hadron Driver for Cancer Therapy 22
Leo Kwee Wah, Ken Takayama, Toshikazu Adachi, Tadamichi Kawakubo, Tanuja Dixit

X-Ray Imaging of an Industrial Microcontroller Sample Using Dual-Energy Technique 34
Azhani Mohd Razali, Roslan Yahya, Maziah Jamaludin

The Impact of Policy Toward R&D and Innovation on Nuclear Technology in Malaysia 39
Fairuz Suzana Mohd Chachuli, Faridah Mohamad Idris

Effect of Particle Size Analysis on the Physicochemical Characterization of Malaysian Monazite 51
Nurliana Roslan, Cik Rohaida Che Hak, Abdul Rahman Norazumin

Feasibility Study of Phosphor Material Synthesis Process Based on Barium Halides Via Hydrothermal Technique 60
Izura Izzuddin, Siti Aishah Ahmad Fuzi, Sapizah Rahim

Effects of Gamma Radiation on Sheep Wool Fiber Composites Reinforced with Rice Husk Ash 65
Norsyahidah Mohd Hidzir, Afifah Mardhiah Mohamed Radzi, Nur Ain Mohd Radzali

Gamma Radiation Shielding Simulation for Thermoplastic Elastomer and Silicon Rubber/Alloy Dopant with Functional Nanoparticle 70
Chai Chan Ful, Muhammad Arif Sazali, Khairulnadzmi Jamaluddin, Nur Syazwani Mohd Ali, Muhammad Syahir Sarkawi

Preliminary Study on Radiation Shielding Properties of Sulfur Concrete on Mars 82
Muhammad Nazrif Zamani, Muhammad Syahir Sarkawi, Asyraf Arif Abu Bakar, Sapizah Rahim, Mohamad Shazwan Ahmad Shah, Sarehati Umar, Norhazilan Md Noor

Attenuation Range of Particle Radiations (alpha and Beta) in Human Skin by Monte Carlo Simulation 89
Nor Alia Aqilah Nawawi, Suffian Mohamad Tajudin, Fairuz Mohd Nasir, Nurul Syazwina Mohamed, Nazifah Abdullah, Nur Hamizah Mohd Zainudin, Juliana Mohd Radzi

Effects of Gamma Rays on Normal-High-Density Concrete 96
Farah Ainin Sofiya Muhd Syazli, Mohd Idzat Idris

Radiation Dosimetry for Quality Control of Silicon Wafer Using Electron Beam 100
Shalina Sheik Muhamad, Siti Zulaiha Hairaldin, Muhd Izham Ahmad, Shahrina Akma Mansur, Noor Hasni M. Ali

Fire Probabilistic Safety Assessment of Control Room in TRIGA PUSPATI Research Reactor...................... 106
Zaredah Hashim, Nik Qurrotu Ain Ahmad Nasruddin, Mohd Huzair Hussain, Faizal Kunchi Mohamed, Phongsakorn Prak Tom, Julia Abdul Karim, Mazleha Maskin

Assessment of Elemental in Plants Acted as Pollution Biological Indicator..114
Zalina Laili, Lakam Mejus, Azian Hashim, Nurul Fairuz Diyana, Mohammad Syahiran Mustaffa, Muhammad Azfar Azman, Rahman Yaccup, Mohd Muzamil Mohd Hashim, Siti Aminah Omar, Shakirah Abd. Shukor, Wilfred Paulus

Assessment and Sources Identification of Air Quality Pollution in Klang Valley, Kuala Lumpur, Malaysia 129
Md Suhaimi Elias, Azian Hashim, Nurul Fairuz Dyana Bahrudin, Nur Aqilah Sapiee, Wilfred Paulus, Muhammad Azfar Azman, Nor Jawahir Raduian, Intan Masayu Abdullah

Potential of Tropical Expansive Clays as Natural Sealing Layer in Nuclear Waste Repository System: A Preliminary Study.. 148
A S N Shah, M A Zali

Author Index

IOP Conference Series: Materials Science and Engineering

International Nuclear Science, Technology and Engineering Conference 2022 (iNuSTEC2022)

"Nuclear Science and Technology Contribution Towards SDG"

Universiti Kebangsaan Malaysia
Bangi, Selangor, Malaysia
25-27 October 2022

Editors

Faridah Mohamad Idris, Malaysian Nuclear Agency
Julie Andrianny Murshidi, Malaysian Nuclear Agency
Faizal Mohamed, Universiti Kebangsaan Malaysia
Irman Abdul Rahman, Universiti Kebangsaan Malaysia
Azman Jalar, Universiti Kebangsaan Malaysia
Norsyahidah Mohd Hidzir, Universiti Kebangsaan Malaysia
Syazwani Mohd Fadzil, Universiti Kebangsaan Malaysia
Nurul Nafiqah Itam Sulaiman, Universiti Kebangsaan Malaysia
Mohd Idzat Idris, Universiti Kebangsaan Malaysia
Aznan Fazli Ismail, Universiti Kebangsaan Malaysia
Imran Yusuff, Universiti Kebangsaan Malaysia
Khoo Kok Siong, Universiti Kebangsaan Malaysia
Hoh Siew Yan, , Universiti Kebangsaan Malaysia
Hassan Mohamed, Universiti Tenaga Nasional
Abdul Aziz Mohamed, Malaysian Nuclear Society

All papers have been peer-reviewed

Supporting/Sponsoring Organisations

Ministry of Education Malaysia; Ministry of Science, Technology and Innovation Malaysia; Ministry of Energy and Natural Resources; Malaysian Nuclear Agency; ASME Malaysia Section; Malaysian Welding and Joining Society; Reactor Interest Group; Asia-Oceania Neutron Scattering Association, Japan Atomic Energy Society; Advanced Nuclear Engineering Research Group

IOP Conf. Series: Materials Science and Engineering 1285 (2023) 011001 doi:10.1088/1757-899X/1285/1/011001

PREFACE

Universiti Kebangsaan Malaysia (UKM) is honoured to host the International Nuclear Science,Technology and Engineering Conference 2022 (iNuSTEC2022). With the theme "Nuclear Application for the Success of the Sustainable Development Goals, SDGs", iNuSTEC2022 is the twelfth in a series of conferences hosted by the Malaysian Nuclear Society (MNS) since 2010, through various research institutes and local universities.

The support from nuclear science committees from Japan, South Korea, China, Thailand, India, Indonesia, United States as well as the International Atomic Energy Agency (IAEA) has taken iNuSTEC to the intenational level. I believe this assembly of researchers will open doors to create more research understandings and collaborations through the participation by scholars and practitioners from universities, research institutions, agencies and industry.

Congratulations to the organizing committee for hosting this iNuSTEC2022, together with Nuclear Youth Competition 2022 (NYC2022). My deepest appreciation goes to our keynote speakers, technical co-sponsor and all supporters for their contribution to ensuring the success of this conference. So, on behalf of UKM, I thank you all, and send best wishes for a very successful conference.

Thank you.

Prof. Dato' Gs. Ts. Dr. Mohd Ekhwan Hj. Toriman
Vice-Chancellor
Universiti Kebangsaan Malaysia
Patron of iNuSTEC2022

IOP Conf. Series: Materials Science and Engineering 1285 (2023) 011001 doi:10.1088/1757-899X/1285/1/011001

MESSAGE FROM:
PRESIDENT, MALAYSIAN NUCLEAR SOCIETY (MNS)

I am really pleased to welcome all keynote speakers, sponsors, presenters, and participants to the International Nuclear Science, Technology and Engineering Conference (iNuSTEC2022). Ideally, the iNuSTEC2022 is organized to enhance the importance of nuclear science, technology, and engineering in developing socio-economic in Malaysia. Malaysian Nuclear Society (MNS) is truly honored to organize this event for the third time with Universiti Kebangsaan Malaysia (UKM). Successful conferences from previous years, at both national and international levels, have brought this event to be more meaningful to strengthen nuclear development for global growth under challenging situation

With the support from the American Society of Mechanical Engineers (ASME) Malaysia Section, and Malaysian Welding and Joining Society (MWJS), the MNS is indeed honored to co-organise this event with UKM and with organizations such as Nuclear Malaysia Agency, UTM, UNITEN, UM, UiTM, USIM and UTHM. The affiliation of the Nuclear Science Club (KESAN) of UKM and Nuclear Engineering Students Society (NESS) of UTM and Institute of Nuclear Engineers (INE) of MNS, to make iNuSTEC2022 a success is a good testament that nuclear still remain relevant in the heart of the young generation to move towards greater strength in nuclear fields. In the future, MNS members shall strive to reach out to more diverse population in order to make nuclear ecosystem more conducive and stronger.

This conference will be participated by experienced and young researchers that can give a platform for them to discuss and debate on the latest development in all areas of nuclear technology applications for friendly purposes at the international level. This would spur and catalyse further research and development leading to the sustainability and improvement of the quality of life without compromising the quality of the environment in Malaysia and the world.

MNS is a unique non-governmental organisation that is formed and headed by individuals that are aligned to the beneficial usage of nuclear science, engineering, and technology which has been incorporated under the Malaysian Society Act. The MNS has been actively promoting the information, advancement, popularization, and understanding of nuclear the public in Malaysia and this conference is one of the ways of realizing such aim. This conference is also served as a public acceptance program to introduce the benefits and advantages of the use of nuclear science, engineering, and technology to the public on a large scale. On top of that MNS also published twice a year on regular basis the Journal of Nuclear and Related Technologies (JNRT) and Buletin Nuklear Malaysia (BNM) for researchers to share their research findings with the community. We also publish iNuSTEC proceedings on yearly basis.

I would like to express gratitude to UKM for hosting iNuSTEC2022, Radiation Protection for Young Professionals, Nuclear Youth Competition 2022 and Post Conference Workshop. I wish to express my heartful thanks to all speakers for your willingness to share your experience with the participants and presenters for the development of the nuclear field. The contribution is not only coming from the Malaysian scientific community but also distinguished guest speakers of iNuSTEC2022. to all the personnel involved for their unwavering efforts to make this event a success. May all these efforts bring benefits to all of us and I congratulate everyone involved in making this conference successful event..

Thank you.

Associate Professor Dr Abdul Aziz Mohamed
President
Malaysian Nuclear Society

CONTENTS

Preface from the Patron of iNuSTEC2022

Message from the President, Malaysian Nuclear Society (MNS)

PAPERS

a) **Science and Engineering**
Paper 1 – 5

b) **Material and Sustainability**
Paper 6 – 13

c) **Environment and Safety**
Paper 14 – 18

IOP Conf. Series: Materials Science and Engineering 1285 (2023) 011001 doi:10.1088/1757-899X/1285/1/011001

iNuSTEC2022 ORGANISING COMMITTEE

Patron
Prof. Dato' Gs. Ts. Dr. Mohd Ekhwan Hj. Toriman
Vice Chancellor, Universiti Kebangsaan Malaysia

Advisor:
Prof. ChM. Dr. Ishak Ahmad,
Dean Faculty of Science and Technology, Universiti Kebangsaan Malaysia

Chairman
Dr. Hoh Siew Yan
Universiti Kebangsaan Malaysia

Co-Chairman
Associate Prof Dr Aziz Mohamed
President, Malaysian Nuclear Society (MNS)

Prof. Dr. Azman Jalar
Chair Department of Applied Physics, Faculty of Science and Technology, Universiti Kebangsaan Malaysia

Deputy Chairman
Assoc. Prof. Dr. Irman Abdul Rahman
Universiti Kebangsaan Malaysia

Co-Deputy Chairman
Prof. Ts. Dr. Faizal Mohamed
Universiti Kebangsaan Malaysia

Secretary
Dr. Nurul Nafiqah Itam Sulaiman
Universiti Kebangsaan Malaysia

Dr. Faridah Mohamad Idris
Secretary, MNS

Treasurer
Dr. Julie Andrianny Murshidi
Treasurer, MNS

TECHNICAL COMMITTEE

Technical and Logistics Committee
Dr. Imran Yusuff, *UKM*
Dr. Hassan Mohamed, *UNITEN*
Cik Siti Hamizah Samsudin, *UKM*
En. Ahmad Takim Saring, *UKM*
En. Che Mohd Zaidi Che Daud, *UKM*
En. Mohd Edzrul Fadly Mat Nawi, *UKM*
En. Mohd Faridz Abd. Jalil, *UKM*
En. Mohd Khairul Azmi Mohd Zuki, *UKM*

Scientific Committee
Dr. Aznan Fazli Ismail, *UKM*
Dr. Mohd Syukri Yahya, *UNITEN*
Dr. Mohd Idzat Idris, *UKM*

Protocol and Publicity Committee
Assoc. Prof. Dr. Khoo Kok Siong, *UKM*
En. Tazidi Taib, *UKM*
En. Azlan Shah Mustafa, *UKM*

Banquets and Souvenirs Committee
Ts. Dr. Norsyahidah Mohd Hidzir
Ts. ChM. Dr. Syazwani Mohd. Fadzil
Pn. Munira Hasan

RADIATION PROTECTION FOR YOUNG PROFESSIONALS SECRETARIAT

Asyraf Arif Bin Abu Bakar, INE MNS/Nuklear Malaysia
Khairul Anwar Bin Bapujee, INE MNS/Nuklear Malaysia
Puteri Nuraliah Husna Binti Mohd Tajuddin, INE MNS/Nuklear Malaysia
Dr Muhammad Syahir Sarkawi, INE MNS/UTM
Muhammad Zulhelmi Bin Mahadi, INE MNS/Nuklear Malaysia
Dr Muhammad Arif Sazali, INE MNS/UTM
Dr. Khairulnadzmi Jamaluddin, INE MNS/UTM

IOP Conf. Series: Materials Science and Engineering 1285 (2023) 011001 doi:10.1088/1757-899X/1285/1/011001

NUCLEAR YOUTH COMPETITION 2022 SECRETARIAT

Project Leader:
Muhammad Hazim Yahya, *KESAN UKM*

Deputy Project Leader:
Farah Hanim Syahirah Ayob, *KESAN UKM*
Mohamad Rizuwan Haikal Abdul Halim, *KESAN UKM*

Secretary:
Dayini Sabrina Zuber, *KESAN UKM*

Co-Secretary:
Nur Farah Najwa Ramli, *KESAN UKM*

Treasurer:
Maizatul Amira Muhamad, *KESAN UKM*

Co-Treasurer:
Sufiah Azman, *KESAN UKM*

Committee Members
Students of *KESAN UKM*

Judges
Dr. Siti Najila Mohd Janib, *Nuklear Malaysia*
En. Azraf Azman, *Nuklear Malaysia*
Muhammad Iman Azhar, *UKM*
Nur Batrisyia Hannani Abdul Rahman, *UKM*

POST CONFERENCE WORKSHOP SECRETARIAT

Dr. Hoh Siew Yan, *UKM*
Dr. Nurul Nafiqah Itam Sulaiman, *UKM*
Pn. Munira Hasan, *UKM*
Cik Siti Hamizah Samsudin, *UKM*
En. Azlan Shah Mustafa, *UKM*
En. Ahmad Takim Saring, *UKM*

LIVE STREAMING SECRETARIAT

En. Azlan Shah Mustafa, *UKM*
En. Mohd Edzrul Fadly Mat Nawi, *UKM*
En. Mohd Faridz Abd. Jalil, *UKM*
En. Mohd Khairul Azmi Mohd Zuki, *UKM*

IOP Conf. Series: Materials Science and Engineering 1285 (2023) 011001 doi:10.1088/1757-899X/1285/1/011001

iNUSTEC 2022 AWARDS AND RECOGNITION

Congratulations All Winners

BEST POSTER AWARD

Nazaratul Ashifa Abdullah Salim, *Nuklear Malaysia*
"Multi Element Assessment in Indegenous Rice and Inter Relationship to Soil of Origin"

Aznan Fazli Ismail, *UKM*
"Rapid Selective Removal of Thorium via Electrosorption Towards Efficiently Managing Rare Earth Extraction Residue"

Fazliana Mohd Saaya, *Nuklear Malaysia*
"Plant Based Radioprotectant as Agent to Expand the Boundaries of Radioprotective Efficacy"

NUCLEAR YOUTH COMPETITION 2022

Champion:
Universiti Sains Islam Malaysia (USIM)
Muhammad Feidhul Hakim bin Fatah Yasin
Abdul Basith bin Abd Samad

2nd Place:
Universiti Tenaga Nasional (UNITEN)
Mohamad Luqman Hafiz bin Md Hanif
Syabil Fikri bin Sabri

3rd Place:
Universiti Malaya (UM)
Jazman Faiz bin Baharuddin
Vincent Tee Ming Han

ACKNOWLEDGEMENT

iNuSTEC2022 is jointly organized by Malaysian Nuclear Society (MNS)
and Universiti Kebangsaan Malaysia (UKM)

In Collaboration with

Malaysian Nuclear Agency (Nuklear Malaysia)
Universiti Teknologi Malaysia (UTM)
Universiti Tenaga Nasional (UNITEN)
Universiti Teknologi MARA (UiTM)
Universiti Islam Antarabangsa Malaysia (IIUM)
Universiti Malaya (UM)
Universiti Sains Islam Malaysia (USIM)
Universiti Tun Hussien Onn Malaysia (UTHM)

Co-organized by
Kelab Sains Nuklear (KESAN)
Nuclear Engineering Students Society (NESS)
Institute of Nuclear Engineers (INE MNS)
Malaysian Nuclear Professional Chapter (MNPC MNS)

Supported by
ASME Malaysia Section
Malaysian Welding and Joining Society

IOP Conf. Series: Materials Science and Engineering 1285 (2023) 011001 doi:10.1088/1757-899X/1285/1/011001

MNS COUNCIL MEMBERS 2021/2023

Advisor / Past President	Prof. Dr. Nahrul Khair Bin Alang Md Rashid (*UTM*)
President	Dr. Abdul Aziz Bin Mohamed (*UNITEN*)
Vice President	Assoc. Prof. Dr. Faizal Bin K.P. Kunchi Mohammed (*UKM*)
Secretary	Dr. Faridah Bt Mohamad Idris (*Nuklear Malaysia*)
Deputy Secretary	Anas Bin Mohamad Fauzi (*UNITEN*)
Treasurer	Dr. Julie Andrianny Bt. Murshidi (*Nuklear Malaysia*)
Deputy Treasurer	Khair'ah Binti Yazid@Khalid (*Nuklear Malaysia*)

Members
Ts. Dr. Hassan Bin Mohamed (*UNITEN*)
Dr. Julia Bt. Abdul Karim (*Nuklear Malaysia*)
Dr. Shukri Mohd (*Nuklear Malaysia*)
Dr. Muhammad Rawi B. Mohamed Zin (*Nuklear Malaysia*)
Dr. Mohd Syukri Yahya (*UNITEN*)

Auditors
Khairul Azhar Bin Abdul Halim (*Nuklear Malaysia*)
Ikmal Azrai Bin Abdul Aziz (*Arey Solutions*)

MNS Chapters
RadTech (*Radiation Tech*)
XApp (*X-ray App*)
NuEnergy (*Nuclear Energy*)
Neutron Photon (*Neutron- Photon*)
Student Chapter (*UKM, UiTM, UNITEN, UTM - NESS*)
Women in Nuclear, Malaysia (*WiN Malaysia*)
Institute of Nuclear Engineers (*INE MNS*)
Malaysian Nuclear Professional Chapter (*MNPC MNS*)

MNPC-MNS COMMITTEE MEMBERS 2022/2023
Advisor / President MNS: Dr. Abdul Aziz Bin Mohamed
President: Ts. Razali Bin Harun (*REM*)
Vice President: Khairul Izzat Bin Abdul Aziz (*Reltech Lab*)
Secretary: Mohd Taha Bin Hj. Suja' (*REM*)

INE MNS COMMITTEE MEMBERS 2022/2023
President: Asyraf Arif Bin Abu Bakar (Nuklear Malaysia)
Vice President: Khairul Anwar Bin Bapujee (Nuklear Malaysia)
Secretary: Puteri Nuraliah Husna Binti Mohd Tajuddin (Nuklear Malaysia)
Comittee: Dr Muhammad Syahir Sarkawi (UTM),
Muhammad Zulhelmi Bin Mahadi (Nuklear Malaysia)

Visit us at http://www.nuklearmalaysia.org

IOP Conf. Series: Materials Science and Engineering 1285 (2023) 011002 doi:10.1088/1757-899X/1285/1/011002

Peer Review Statement

All papers published in this volume have been reviewed through processes administered by the Editors. Reviews were conducted by expert referees to the professional and scientific standards expected of a proceedings journal published by IOP Publishing.

1. Type of peer review:

2. Conference submission management system: Morressier

3. Number of submissions received: 20

4. Number of submissions sent for review: 20

5. Number of submissions accepted: 18

6. Acceptance Rate (Submissions Accepted / Submissions Received × 100): 90 %

7. Average number of reviews per paper: 1.38

8. Total number of reviewers involved: 21

9. Contact person for queries:

Name: Faridah Mohamad Idris

Affiliation: Malaysian Nuclear Society (MNS)

Email: faridah_idris@nm.gov.my

* means value has been edited

IOP Conf. Series: Materials Science and Engineering 1285 (2023) 012001 doi:10.1088/1757-899X/1285/1/012001

Pragmatic modelling of axial flow-induced vibration (FIV) for nuclear fuel rods

Anas Muhamad Pauzi[1,2,a)], Hector Iacovides[1] and Andrea Cioncolini[3]

[1]Dept. of Mech, Aero. and Civil Eng. (MACE), University of Manchester, M13 9PL Manchester, UK
[2]College of Engineering, Universiti Tenaga Nasional, Jalan IKRAM-UNITEN, 43000 Kajang, Malaysia
[3]Guangdong Technion-Israel Institute of Technology [GTIIT], Shantou, Guangdong, China

[a]anas@uniten.edu.my/ anaspauzi@gmail.com

Abstract. Flow-induced vibration (FIV) at the spacer grid in the fuel assembly of a Light Water Reactor (LWR) is the leading cause of fuel failure. This project aims to produce a simulation benchmark on the experimental campaign at MACE on axial FIV of a cantilever beam in an annular tube, that mimics the configuration and environment of a typical LWR. The nuclear fuel rod, which consists of fuel pellets filled in Zirconium alloy cladding is modelled in the experiment as a steel rod filled with lead shots that closely approximates the filling density of the fuel pellet. To reduce the complexity and increase the efficiency of the simulation, further simplification is applied by on the geometry and assuming the solid domain as a single material instead of multiple materials. The first mode of frequency of the rod vibrating in quiescent water, which had been validated via the Euler-Bernoulli beam theory against experimental measurements, was used to design the single-material solid domain. Two models were proposed, firstly a solid rod with lower density and stiffness (SLE), and secondly an empty cladding with high density and low stiffness (EHD). Both solid and fluid domains were discretised using the cell-centred finite volume (FV) method and coupled with strong two-way fluid-structure interaction (FSI). Results on the frequency of vibration in quiescent water and in axial flow showed good agreement with experimental measurement, and the computational efficiency is analyzed for different rod models and changes in parameters of the solid domains.

1. Introduction

The flow-induced vibration (FIV) in the nuclear fuel rods, which leads to fretting wear at the grid spacer were the main cause of fuel failure in the PWR reactor [1]. The fuel rods were loosely assembled with springs at both ends and positioned with spacer grids consisting of springs and dimples designed to allow for the thermal expansion of fuel rods during operation but, this had led to continuous FIV, hence the fretting wear. The project aims to benchmark the experiment on the axial water flow in a tube pipe towards a cantilever beam, where measurements involve validating both flow fields and their corresponding vibration response [2]. A pragmatic approach was used to achieve an optimal balance between the accuracy of the results and computation efficiency.

2. Literature review

To establish a better understanding of the dynamics of the flow-induced vibration for nuclear applications, most earlier studies focused on flexible rods which result in larger displacements [3]. Meanwhile, to reduce complexity, the nuclear fuel rod is commonly simplified as a single material [3], while others modelled the rod with solid lead-bismuth at the centre to represent the fuel pellets [4].

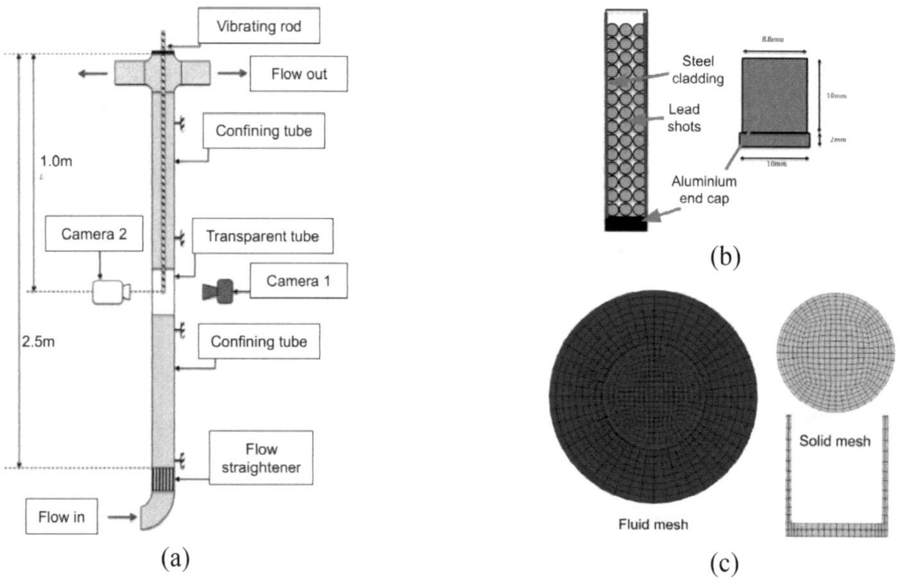

Figure 1. MACE axial FIV schematic of (a) experimental setup, (b) lead-filled rod, and (c) XY-plane cross-section of the fluid (left), solid (top-right) and XZ-plane near free for EHD rod (bottom right)

The MACE FIV experiments, on the other hand, shown in figure 1 (a) and (b) used variable-sized lead spheres filled in a steel cladding with similar packing factor as the uranium fuel pellet used in the nuclear fuel rod [2], and the results on the 1st mode of vibration proved good agreement with the analytical Euler-Bernoulli beam theory, by assuming that the flexural rigidity is calculated based on the geometry of the cladding only, neglecting the effects of the lead-spheres. Previous benchmark studies performed at Reynolds number 16,400 attained excellent accuracy using the rod modelled as a solid single material cylinder [5]. Preliminary simulations on the axial FIV involving a higher Reynolds number showed an increase in tenfold computation requirement as the Reynolds number doubled. Other methods used to increase computational efficiency such as applying the Finite Element Method (FEM) to solve for the solid domain, which resulted in faster computation as compared to the Finite Volume Method (FVM) for steel beam vibration in an axial water flow test but added extra interpolation at the FSI interface [6]. Some studies also restrict simulation only on the fluid domain to obtain the pressure fluctuation near the rod and then used to predict the corresponding vibration using an analytical forced vibration model [2,7]. This method would neglect the effects of the 2-way FSI.

3. Methodology

3.1. Euler-Bernoulli beam theory

The simplification of the lead-filled steel rod into a single material was performed using the Beam-Bernoulli theory to achieve a similar prediction for the 1st mode of frequency, f_1.

$$f_1 = \frac{\beta_1^2}{2\pi}\sqrt{\frac{EI}{(\dot{m}_{rod} + \dot{m}_{add})\, L^4}} \quad (1)$$

The influence factor, β of the 1st mode of vibration of a fixed-free end is 1.875 [2], E is Young's Modulus, I_{rod} is the 2nd moment of inertia, m_{rod} and m_{add} are the linear mass density of the rod and the added mass, given equation (2), (3), and (4).

$$I_{rod} = \frac{\pi}{4}\left(r_{ro}^4 - r_{ri}^4\right) \quad (2)$$

$$\dot{m}_{rod} = \rho_s \frac{\pi}{4}\left(d_{ro}^4 - d_{ri}^4\right) \quad (3)$$

$$\dot{m}_{add} = \rho_l \frac{\pi}{4} d_{ro}^2 \frac{1 + (d_{ro}/d_{ti})^2}{1 - (d_{ro}/d_{ti})^2} \quad (4)$$

Equations (1) to (4) showed that the density of the rod is inversely proportional to the frequency of vibration and proportional to the elasticity and cross-section area of the rod. With constant 1st mode of vibration, two solid models were proposed, the solid cylinder low elasticity (SLE) and the empty cladding high density (EHD). The comparison with the MACE experiments is tabulated as follows.

Table 1. Geometrical and material properties of the cantilever rod used for the axial FIV experiment and the pragmatic models.

Cantilever rod model	**MACE**	**SLE**	**EHD**
Tube internal diameter, d_{ti} [mm]	21.0	21.0	
Rod outer diameter, d_{ro} [mm]	10.01	10.01	
Length of rod, L_{rod} [mm]	1060	1060	
Rod inner diameter, d_{ri} [mm]	8.83	0	8.83
2nd moment of inertia, I [mm^4] from Eq. (4)	194.4	492.8	194.4
Young's modulus, E [GPa]	193	79.76	202.26
Density of rod, ρ_s [kg/m^3] from Eq. (3)	7990	7471.7	33676.3
Flexural rigidity, EI [N.m^2] from Eq. (1)	37.53	39.33	
Linear mass density of rod, $\dot{m}_{rod}$ [kg/m] from literature [1,2]	0.588	0.588	
Linear mass density of added mass, $\dot{m}_{add}$ [kg/m] from Eq. (2)	0.1244	0.1244	
1st mode of frequency in quiescent water, f_l[Hz] from Eq. (1)	3.61	3.7	
1st mode of frequency in quiescent water, f_l[Hz] from exp. [1]	3.7	-	
Natural frequency in vacuum, f_0 [Hz] from Eq. (1)	4.07	4.07	

From Table 1, the predicted 1st mode of frequency for MACE lead-filled rod is slightly lower compared to the experiment and was expected due to the slightly lower flexural rigidity calculated from the geometry of the steel cladding. This gap in rigidity comes from the contribution of the lead spheres filled in the steel cladding, which is approximately 5%.

Table 2. Geometrical and material properties near the free end, and pragmatic modelling.

	Steel	**Aluminium**	**EHD-rod**
Density, ρ_s [g/mm^3]	7.99	2.74	33.676
Volume, Vol [mm^3]	174.6	765.6	103.8
Mass, m [g]	1.4	2.1	3.5
Length at cap, L [mm]	10	12	1.32

Since the EHD-rod consists of very high mass concentrated at the cladding, an accurate thickness of the free-end cap needs to be calculated based on the measured weight of the aluminium end cap and the cladding surrounding the cap. The length of the EHD rod solid cap is calculated from the total mass of the solid near the free end given and related details tabulated in table 2.

3.2. Finite Volume Solid Deformation

Both solid and fluid domains were solved in segregation using software "Foam-Extend-4.0", and an extension "Solids4Foam" added to facilitate the FSI coupling. The interface quasi-Newton with inverse Jacobian from a least-squares model (IQN-ILS) algorithm, which proved efficient for steel-water coupling is used for the FSI coupling [8]. In a 2-way FSI as summarized in figure 2, at every time-step, the residual, or the gap between the fluid and solid mesh at the interface would be calculated, and if the value falls below the convergence criteria, the residual will be recalculated in the FSI iteration loop. In each FSI iteration, the traction force from the fluid domain is obtained by solving the mesh motion and fluid equations, and subsequently updating the traction force at the interface to solve for the new solid displacement, hence the new residual. Preliminary simulation test showed that the backlog comes from obtaining convergence in the solid solver, hence, it is important to optimize the solid solver to achieve significant computation speed-up.

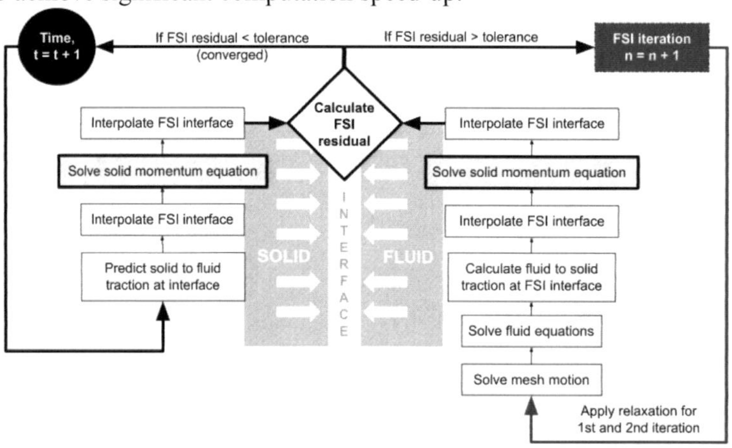

Figure 2: Flow chart for 2-way strong FSI coupling

The governing equation for solid deformation, D, is derived from the first principle using the force-stress equilibrium given in equation (5).

$$\rho_s \frac{\partial^2 D_i}{\partial t^2} + \frac{\partial \sigma_{ij}}{\partial x_j} + f_I = 0 \tag{5}$$

The solid domain was assumed to be isometric, with small strain and rotation neglected. Hence, the small-strain assumption was used for the strain-deformation relation in equation (6) and the stress-strain relation is modelled using the Hookean model as in equation (7).

$$\epsilon_{ij} = \frac{1}{2}\left(\frac{\partial D_i}{\partial x_j} + \frac{\partial D_j}{\partial x_i}\right) \tag{6}$$

$$\sigma_{ij} = 2\mu\epsilon_{ij} + \lambda\delta_{ij}\epsilon_{kk} \tag{7}$$

λ and μ are the first and the second Lamé's constant related to Young's modulus and Poisson's ratio of a material, given in equation (8).

$$\lambda = \frac{\nu E}{(1+\nu)(1-2\nu)}, \mu = \frac{E}{2(1+\nu)} \tag{8}$$

The governing equation of motion for solid deformation is derived by substituting equations (6) and (7) into the force-deformation in equation (5), hence becoming as follows.

$$\rho_s \frac{\partial^2 D_i}{\partial t^2} + [\lambda + \mu]\frac{\partial}{\partial x_i}\left(\frac{\partial D_j}{\partial x_j}\right) + \mu\frac{\partial^2 D_i}{\partial x_j^2} + \rho_s f_i = 0 \tag{9}$$

To ensure convergence, the governing equation in equation (9) is decomposed into two parts, where the second term is solved implicitly, while the third term which forms a higher computational pure rotation, would be solved efficiently with explicit treatment, as in equation (10) below [9].

$$\rho_s \frac{\partial^2 D_i}{\partial t^2} - [2\mu + \lambda]\frac{\partial}{\partial x_i}\left(\frac{\partial D_j}{\partial x_j}\right) - \frac{\partial}{\partial x_i}\left(\mu\frac{\partial D_i}{\partial x_j} + \lambda\delta_{ij}\frac{\partial D_i}{\partial x_i} + (\mu + \lambda)\frac{\partial D_j}{\partial x_j}\right) = -\rho_s f_i \tag{10}$$

Meanwhile, the governing equation for the fluid flow follows the incompressible Navier-Stoke equations with Arbitrary Lagrangian-Eulerian (ALE) formulation for mesh motion, given as follows.

$$\frac{\partial U_i}{\partial x_i} = 0, \qquad \frac{\partial U_i}{\partial t} + \frac{\partial}{\partial x_j}\left[(U_i - (U_s)_i)U_j\right] = -\frac{1}{\rho_f}\frac{\partial P}{\partial x_i} + \nu \frac{\partial}{\partial x_j}\left(\frac{\partial U_i}{\partial x_j}\right) \tag{11}$$

The ALE formulation obeys the Geometry Conservation Law (GCL), where the rate of change of volume of a cell is equal to the velocity of the surfaces surrounding the cell, U_s. The fluid mesh deformation was solved using the Laplace equation of motion with varying diffusivity, based on cell distance to the nearest moving boundary, *l*, given in equations (12) and (13) respectively.

$$\frac{\partial}{\partial x_i}\left(\gamma \frac{\partial U_s}{\partial x_j}\right) = 0 \tag{12}$$

$$\gamma(l) = \frac{1}{l^2} \tag{13}$$

The turbulent was modelled using the unsteady Reynolds Average Navier-Stoke (URANS) method, where the Reynolds stresses are solved using the Reynolds Stress Model (RSM), Launder-Reece-Rodi (LRR) model. All time and spatial terms were discretized with second-order schemes, while the cross-section view of the fluid and solid mesh is shown in figure 1 (c) above.

4. Results and discussions

4.1. Free vibration

The free vibration test was conducted on the solid domain only by applying initial traction force and release. Accurate results were obtained by firstly reducing the time-step size to at least 10^{-4}s which was several magnitudes lower than the period vibration, 0.3s. This is required because the transfer of information for solids in the FV method depends on the solid Courant number, which was related to the speed of wave in a solid. Secondly, the mesh at regions of high stresses such as near the fixed end were refined until it reached mesh independence. The HDE rod was approximately 40 times faster computational time than the SLE rod, even though the number of cells is only 3 times higher, hence was used on all subsequent simulations.

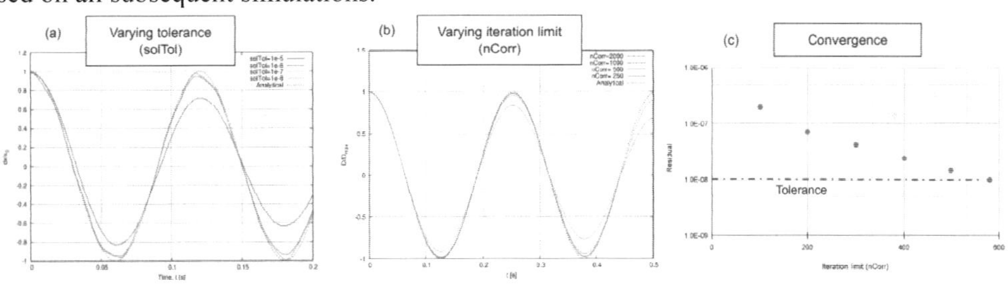

Figure 3. Free vibration of cantilever rod in a vacuum with varying solid solver settings; (a) tolerance and (b) limit number of iteration. (c) The residuals when solver reached iteration limit.

Finally, optimization on the computation time was obtained by varying solid solver settings where numerical damping presents as tolerance and iteration limit decreases as in figure 3 (a) and (b). The latter effectively improves computational time without significantly reducing the accuracy of results. The average residual convergence in a solid solver as shown in figure 3 (c) shows that reducing the number of iterations below 200 could still obtain convergence of 1×10^{-7}.

4.2. Flow-induced vibration

Simulation results of the axial FIV for Re=32,100 in figure (4) proved good agreement with experimental data. The rod vibrates at random with varying amplitude, but the overall amplitude is half of the experimental data. Both frequency spectra gave similar magnitude and trend of the high peak at the 1st and 2nd modes of vibration. Furthermore, reducing the maximum solid solver iteration from 250 to 150 would not reduce the accuracy for both frequency and amplitude of vibration while doubling the computation speed.

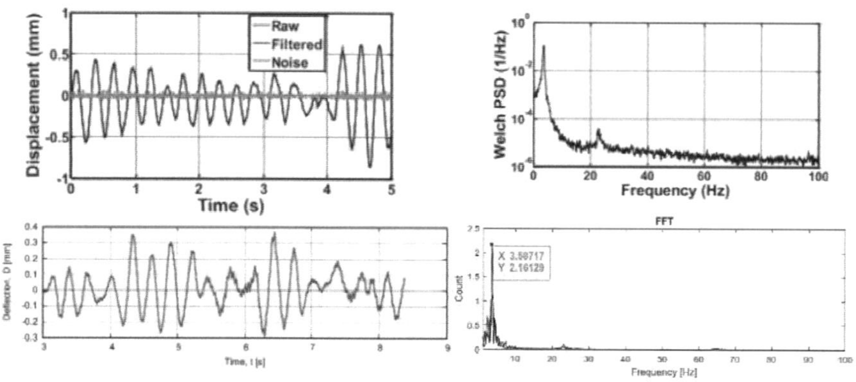

Figure 4. Comparison between (top) experimental [2] and (bottom) simulation results for axial FIV; (left) time-deflection curve, and (right) the FFT frequency spectrum.

5. Conclusions

Good accuracies were obtained for both tests, the free vibration and the axial FIV. The lead-filled steel rod can be modelled as a single material either solid low elasticity (SLE) or empty high density (EHD) rod while the EHD rod gives significant computational speed-up. Significant computation speed-up was obtained by optimizing the solid solver, notably limiting the number of solid solver iterations. Further benchmarking on axial FIV involving higher Reynolds numbers could be performed accurately within available computational resources.

6. References

[1] IAEA 2010 Review of fuel failures in water cooled reactors No.NF-T-2.1.
[2] Cioncolini A, Silva-Leon J, Cooper D, Quinn M K, Iacovides H 2018 Axial-flow-induced vibration experiments on cantilevered rods for nuclear reactor applications *Nuclear Engineering and Design* **338** 102–118.
[3] Paidoussis M 2016 Fluid-structure interactions **2** (Academic Press).
[4] De Santis D, Shams A 2017 Numerical modelling of flow-induced vibration of nuclear fuel rods *Nuclear Engineering and Design* **320** 44–56.
[5] Salachna J, Cioncolini A, Iacovides H 2022 Benchmark Simulation of the flow-induced vibrations for nuclear applications, *SSRN Electronic J.*
[6] ter Hofstede E, Kottapalli S, Shams A 2017 Numerical prediction of FIV in nuclear reactor applications *Nuclear Engineering and Design* **319** 81–90.
[7] Kottapalli S 2017 Numerical prediction of flow-induced vibrations in nuclear reactor applications (TU Delft) .
[8] Degroote J, Bathe K J, Vierendeels J 2009 Performance of a new partitioned procedure versus a monolithic procedure in FSI *Computers and Structures* **87**(11–12), 793–801.
[9] Jasak H, Weller H G 2000 Application of the FVM and unstructured meshes to linear elasticity *Int. J. for Num. Meth. in Eng.* **48**(2), 267–287

Acknowledgement

This publication is part of the postgraduate study sponsored by the Malaysian government via Majlis Amanah Rakyat (MARA) and supported by the National Energy University (UNITEN), Malaysia.

IOP Conf. Series: Materials Science and Engineering 1285 (2023) 012002 doi:10.1088/1757-899X/1285/1/012002

Improving source convergence diagnostics sensitivity using the complex components of Fourier fundamental mode

K F Chin[1], M R Omar[1, a)] and J A Karim[2]

[1] School of Physics, Universiti Sains Malaysia, 11800 USM, Penang, Malaysia.
[2] Reactor Technology Centre, Technical Support Division, Malaysian Nuclear Agency, 43000 Bangi, Selangor, Malaysia.

[a] rabieomar@usm.my

Abstract. Previous works have demonstrated that the Fourier fundamental mode coefficient (FFMC) is a potential fission source convergence indicator for Monte Carlo k-eigenvalue calculations. Unlike the conventional Shannon entropy indicator, it requires no spatial meshing, and its computation is computationally inexpensive. However, we found that FFMC is less sensitive to the convergence involving near symmetrical fission distributions. To rectify the problem, we focus on the analysis of the FFMC's components (the real and imaginary parts) on a simple geometry problem and a full reactor core problem. The usage of the FFMC's components is more conservative compared to the original FFMC method as supported by the improved results of the fission source convergence detection.

1. Introduction

With nowadays high-speed computational power, the Monte Carlo simulation is favoured by researchers and scientists to deal with the nuclear criticality calculation, also known as the k-eigenvalue calculation. However, one of the challenges faced by the simulation in the eigenvalue calculation is the determination of the convergence of the fission source for a particular system which is free from the initial guess distribution. A traditional indicator based on the information theory is borrowed to diagnose the fission source convergence, namely Shannon entropy. This indicator is widely adopted by the community in their codes [1–5]. On the other hand, Omar [6] formulated an indicator based on the Fourier functional expansion to deal with the convergence diagnosis, called the Fourier fundamental mode coefficient (FFMC). The feature of both indicators is that only one single scalar metric is used to represent the N-dimensional distribution quantities. The advantage of the FFMC indicator is that it is mesh-free, which reduces the user's subjectivity to the meshing scheme as required by the Shannon entropy [7].

2. Fission source convergence indicator

2.1. Fourier fundamental mode coefficient (FFMC)

To calculate the Fourier fundamental mode coefficient (FFMC) for a specific system, a cuboid region with volume, $V = L_x L_y L_z$, is defined, called the FFMC box, to enclose the entire fissile region. Then, for each fission generation or MC cycle, the value of FFMC is tallied from the fission site, which is given by:

$$c_\mathrm{f} = \left(c_\mathrm{Re}^2 + c_\mathrm{Im}^2\right)^{1/2} \tag{1}$$

where c_{Re} and c_{Im} are the real and imaginary components of FFMC. At each MC cycle, the values of the components are obtained by [7]:

$$c_{\text{Re}} = \frac{1}{V}\sum_{r\in F} cos(k_{\text{f}} \cdot r) \tag{2}$$

and

$$c_{\text{Im}} = \frac{-1}{V}\sum_{r\in F} sin(k_{\text{f}} \cdot r) \tag{3}$$

where

$$k_{\text{f}} = \left(\frac{2\pi}{L_x}, \frac{2\pi}{L_y}, \frac{2\pi}{L_z}\right) \tag{4}$$

is a fundamental mode vector and F is the entire set of fission site locations generated during each MC cycle. Note that equations (2) and (3) are the result of the Euler's formula based on the derived estimation [6]:

$$\hat{c}_{\text{f}} = \frac{1}{V}\sum_{r\in F} exp[-i(k_{\text{f}} \cdot r)]. \tag{5}$$

The usage of c_{f} in equation (1) only retains the magnitude information without the phase information, which may lead to false convergence determination. To rectify the issue, both component coefficients as stated in equations (2) and (3) are used, so that the whole system information can be reserved, which provides a more conservative result.

2.2. *Shannon entropy*

Shannon entropy is a traditional convergence indicator, which can be calculated by:

$$H = -\sum_{i=1}^{B} f_i log_2 f_i \tag{6}$$

where f_i is the fraction of fission source sites in the i-th mesh bin to the total number of fission source sites generated for each MC cycle. Note that B denotes the total number of mesh bins and is usually declared by the user with a particular meshing scheme.

3. Validation and discussions

The simulation was done by the OpenMC code [3] with modifications to implement the FFMC calculation [6]. To ease the source convergence diagnosis by visualization, the values of c_{f}, c_{Re}, and c_{Im} are normalized to their first cycle values, with the capitalized symbols representing their normalized value, e.g., C_{f}, C_{Re}, and C_{Im}, respectively. Two test cases were simulated in this paper: the fissile columns problem and the BEAVRS (Benchmark for Evaluation and Validation of Reactor Simulations) full core problem.

3.1. *Case I: Fissile columns*

The fissile columns case is a near symmetrical modified from the OECD/NEA source convergence slabs problem [8]. In this case, two uranyl nitrate fissile columns are decoupled by a water column as illustrated in figure 1. The dimension of the problem is 100cm×100cm×100cm, with reflective the boundary condition is imposed for planes normal to y- and z-axes, whereas the vacuum boundary condition is imposed for the x-axis. Two different lengths of fissile columns were used, i.e., 35cm and 36cm, to avoid symmetrical source distribution. The initial distribution is a point source located in the first fissile column at 30cm from the vacuum boundary. The simulation was run for 5,000 and 10,000 neutron histories per cycle.

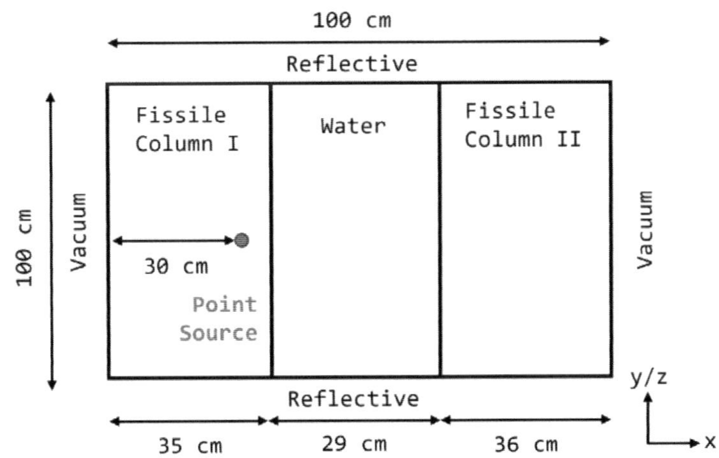

Figure 1. The geometry of the case of the fissile columns. The figure is not drawn to scale.

A 20×20×20 mesh enclosing the entire problem space is used for the entropy calculation, while the dimension of the FFMC box used is 200cm×200cm×200cm for the FFMC calculation. The FFMC box is larger than the problem space to increase the sensitivity of FFMC. The evolution of H, C_f, C_{Re}, and C_{Im} over 5,000 inactive cycles are illustrated in figures 2 and 3 for 5,000 and 10,000 neutron histories per cycle, respectively. For the 5,000 histories case, H, C_f, and C_{Re} reached the equilibrium fluctuations at around 200 MC cycles while C_{Im} reached around 1600 MC cycles. On the other hand, for the 10,000 histories case, similar behaviour is observed for H, C_f, and C_{Re} but C_{Im} reached the equilibrium fluctuations at around 2600th MC cycle. More cycles are needed for the C_{Im} to converge. The squaring operation in equation (1) reduces the convergence effect of C_{Im} from positive to negative values or vice versa, which results in the subtle evolution of C_f toward equilibrium. Figure 4 shows the ratio of sources in fissile column II to the total number of sources. The result indicates that H, C_f, and C_{Re} give false convergence signal as the ratio of source in fissile column II did not reach equilibrium at the 200th MC cycle.

To further illustrate the consequence of premature convergence determined by H, C_f, and C_{Re}, the fission rate of fixed 2,000 active cycles is calculated with 3 different numbers of inactive cycles, i.e., 200, 1,600, and 3,000. The calculation of the fission rate along the x-axis is divided by 20 mesh bins. Figure 3 shows the fission rate with its (3 times standard deviations) error bars. The curves of 200 inactive cycles for both histories clearly show that large divergence appears as compared to the curves of 3,000 inactive cycles. The average relative errors of the fission rate are listed in table 1 with the results of 3,000 inactive cycles as the reference. The results indicate the inactive cycles determined by C_{Im} reduce the error of the fission rate as compared to other indicators. Hence, both the component indicators, i.e., C_{Re} and C_{Im}, are suggested to be used simultaneously for this slow converging problem.

In general, the computational cost of the FFMC method is lower than the Shannon entropy method. In terms of the number of memory allocations, the FFMC method requires two memory allocations for the two tallies c_{Re} and c_{Im} while the Shannon entropy method requires B memory allocations according to equation (6) and is determined by users [7]. On the other hand, the computational time can be calculated from the exported source locations. The exportation is to exclude the simulation time for the neutron events. The results are displayed in table 2. Note that the relative time taken is computed by taking the entropy value as the relative base. In this case, the FFMC method saved around 4% computational time as compared to the Shannon entropy method.

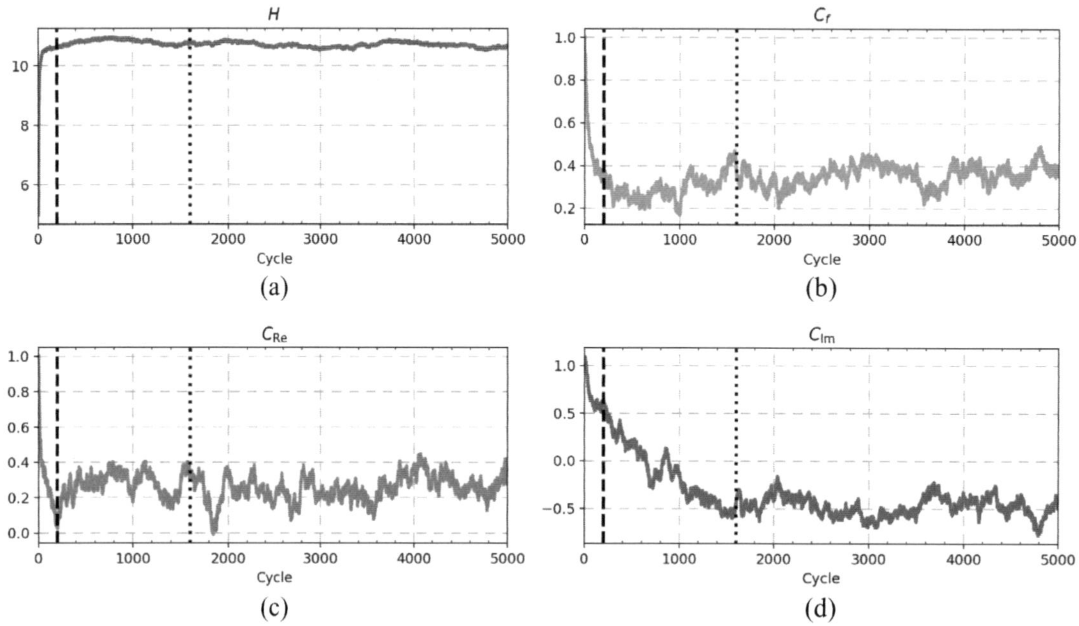

Figure 2. Evolution of the values of (a) Shannon entropy (H), (b) normalized FFMC (C_f), (c) the normalized real component of FFMC (C_{Re}), and (d) the normalized imaginary component of FFMC (C_{Im}) over 5,000 inactive cycles with 5,000 neutron histories per cycle. The dashed line marked the 200th MC cycle while the dotted line marked the 1600th MC cycle.

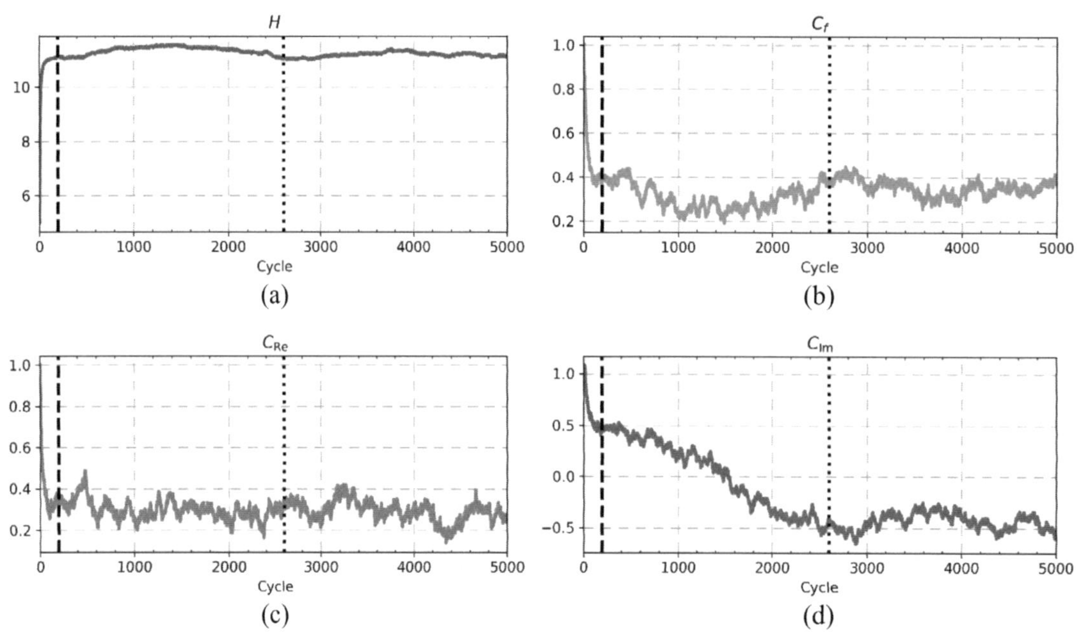

Figure 3. Evolution of the values of (a) Shannon entropy (H), (b) normalized FFMC (C_f), (c) the normalized real component of FFMC (C_{Re}), and (d) the normalized imaginary component of FFMC (C_{Im}) over 5,000 inactive cycles with 10,000 neutron histories per cycle. The dashed line marked the 200th MC cycle while the dotted line marked the 2600th MC cycle.

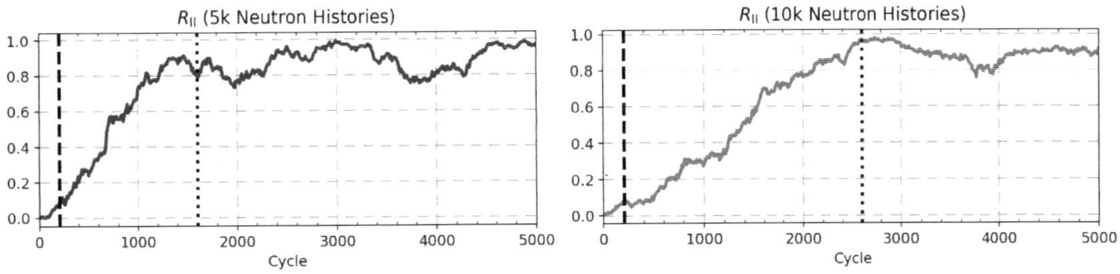

(a) (b)

Figure 4. Evolution of the ratios of source in fissile column II to the total number of sources (R_{II}) over 5,000 MC cycles for (a) 5,000 and (b) 10,000 neutron histories per cycle, respectively. The dashed line marked the 200th MC cycle for both histories while the dotted line marked the 1600th and 2600th MC cycles for 5,000 and 10,000 neutron histories, respectively.

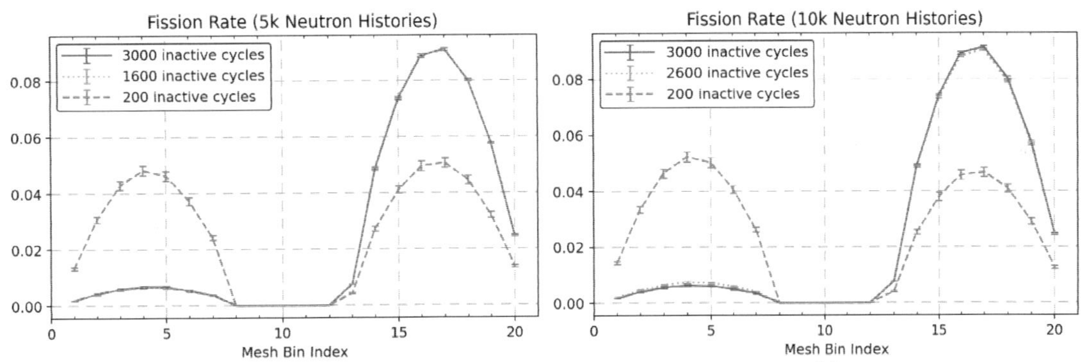

(a) (b)

Figure 5. Fission rates along x-direction with the 3 different numbers of inactive cycles for (a) 5,000 and (b) 10,000 neutron histories per cycle, respectively.

Table 1. Average relative errors of fission rate with different numbers of inactive cycles for 5,000 and 10,000 neutron histories per cycle.

Neutron histories	Number of inactive cycles	Average relative error
5,000	200	257.8%
	1600	1.0%
	3000	-
10,000	200	308.5%
	2600	7.6%
	3000	-

Table 2. Computational time between Shannon entropy and FFMC.

Neutron histories	Relative time taken	
	Shannon entropy	FFMC
5,000	1.0	0.958
10,000	1.0	0.956

IOP Conf. Series: Materials Science and Engineering 1285 (2023) 012002 doi:10.1088/1757-899X/1285/1/012002

3.2. Case II: BEAVRS full core

BEAVRS is a full core PWR benchmark developed by the MIT Computational Reactor Physics Group [9]. The complex geometry of the full core was simulated with 193 fuel assemblies in a 15×15 grid as plotted in figure 6. The initial source distribution is a point source located at the centre of the fissile region. In this case, 100,000 neutron histories per cycle were simulated. A 15×15×20 mesh enclosing the entire fissile region is used for the entropy calculation, while a larger FFMC dimension is used for the FFMC calculation, i.e., 600cm×600cm×550cm. Figure 7 illustrates the evolution of H, C_{f}, C_{Re}, and C_{Im} over 800 inactive cycles. The evolution of H, C_{f}, and C_{Re} achieved equilibrium at around 150th MC cycle while for C_{Im} achieved at around the 600th MC cycle. The convergence of C_{Im} is determined by noticing that there exists a minor decreasing trend at around the 500th MC cycle.

The fission rate had also been calculated for this case. A fixed number of 100 active cycles was used and 3 different numbers of inactive cycles were calculated, i.e., 150, 600, and 700. The same mesh scheme (15×15×20) was used for the fission rate calculation. The average relative errors of the fission rate are listed in table 3, with the results of 700 inactive cycles as the reference. The convergence instance indicated by C_{Im} improved the accuracy of the results.

For the computational cost comparison, the results of the relative computational time taken are presented in table 4. The FFMC method is a better choice of the convergence indicator since it reduced the computational time by 7% as compared to the entropy method. Note that the computational time taken between the calculations of FFMC and the components of FFMC is insignificant because both algorithms only differ by the computation of equation (1).

(a)

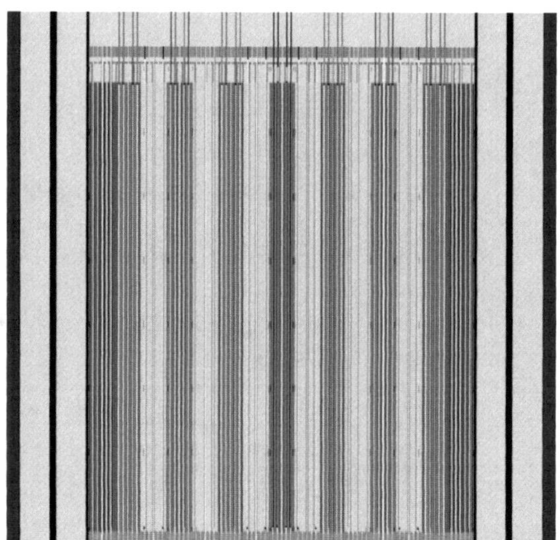
(b)

Figure 6. (a) Radial geometry and (b) axial geometry of the BEAVRS full core case.

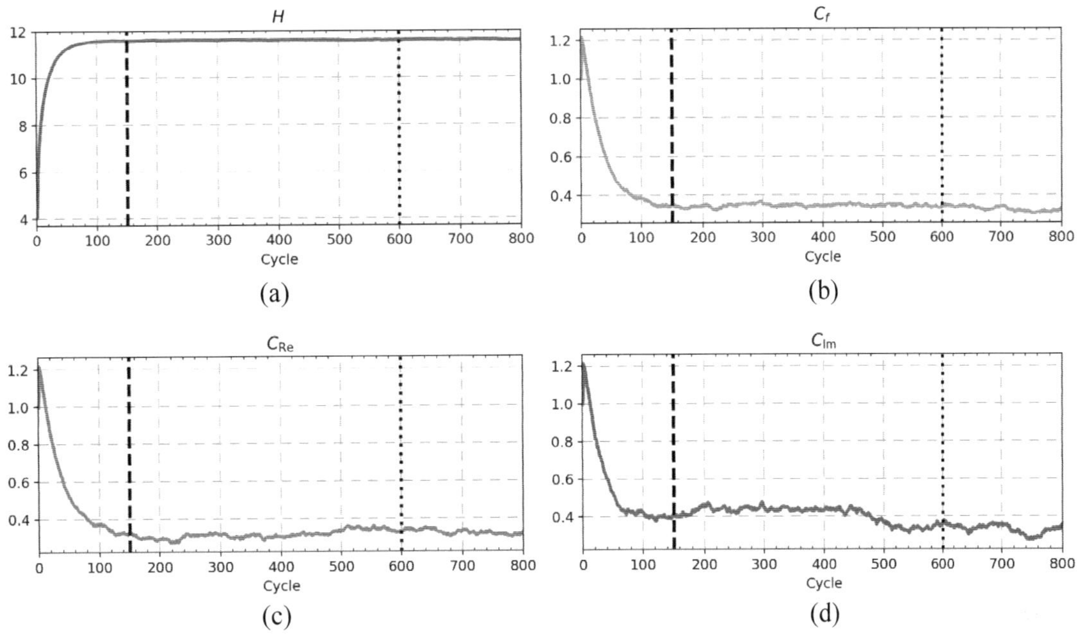

Figure 7. Evolution of the values of (a) Shannon entropy (H), (b) normalized FFMC (C_{f}), (c) the normalized real component of FFMC (C_{Re}), and (d) the normalized imaginary component of FFMC (C_{Im}) over 800 inactive cycles for BEAVRS full core case. The dashed line marked the 150th MC cycle while the dotted line marked the 600th MC cycle.

Table 3. Average relative errors of fission rate with different numbers of inactive cycles for BEAVRS full core case.

Number of inactive cycles	Average relative error
150	8.31%
600	8.12%
700	-

Table 4. Computational cost between Shannon entropy and FFMC for BEAVRS full core case.

	Relative time taken
Shannon entropy	1.0
FFMC	0.926

4. Conclusions

In summary, the FFMC indicator only uses the magnitude information from the Fourier series expansion which can lead to a false convergence judgement as illustrated in this paper. To rectify the issue, the components of FFMC should be used, i.e., the real and imaginary components of FFMC. The validation was done for two test cases: the fissile columns case and the BEAVRS full core case. The numerical results of the fission rate for the two cases had been improved with the convergence judgement of FFMC's components.

5. References

[1] Goorley T *et al* 2016 Features of MCNP6 *Ann. Nucl. Energy* **87** 772–83

[2] Leppänen J, Pusa M, Viitanen T, Valtavirta V and Kaltiaisenaho T 2015 The Serpent Monte Carlo code: Status, development and applications in 2013 *Ann. Nucl. Energy* **82** 142–50

[3] Romano P K and Forget B 2013 The OpenMC Monte Carlo particle transport code *Ann. Nucl. Energy* **51** 274–87

[4] Wang K, Li Z, She D, Liang J, Xu Q, Qiu Y, Yu J, Sun J, Fan X and Yu G 2015 RMC – A Monte Carlo code for reactor core analysis *Ann. Nucl. Energy* **82** 121–9

[5] Omar M R, Karim J A and Yoon T L 2019 The development of a multigroup Monte Carlo code for TRIGA reactors *Nucl. Eng. Des.* **342** 99–114

[6] Omar MR 2021 An effective mesh-free fission source convergence indicator for Monte Carlo *k*-Eigenvalue problems *Nucl. Eng. Des.* **372** 110960

[7] Omar M R and Karim J A 2022 Fission source stationarity diagnostics using the Fourier fundamental mode coefficient *Prog. Nucl. Energy* **146** 104164

[8] Blomquist R, Nouri A, Armishaw M, Jacquet O, Naito Y, Miyoshi Y and Yamamoto T 2003 OECD/NEA source convergence benchmark program: Overview and summary of results *Proc. Int. Conf. Nuclear Criticality Safety (ICNC2003), 20–24 October 2003, Tokai, Ibaraki* pp 278–82

[9] Kumar S, Liang J, Forget B and Smith K 2020 BEAVRS: An integral full core multi-physics PWR benchmark with measurements and uncertainties *Prog. Nucl. Energy* **129** 103488

Acknowledgements

The authors wish to express their gratitude to Universiti Sains Malaysia for financing this research through Short Term Grant, no. 304/PFIZIK/6315550.

Nuclear photonics: Laser-driven nuclear physics

J. F. Ong[1,a)], Meng-Hock Koh [2, 3,b)] and I. H. Hashim[2,3]

[1]Extreme Light Infrastructure - Nuclear Physics (ELI-NP), "Horia Hulubei" National Institute for Physics and Nuclear Engineering (IFIN-HH), 30 Reactorului Street, RO-077125 Bucharest-Măgurele, Romania
[2]Department of Physics, Faculty of Science, Universiti Teknologi Malaysia, 81310 Johor Bahru, Johor, Malaysia
[3]UTM Centre for Industrial and Applied Mathematics, 81310 Johor Bahru, Johor, Malaysia

[a]jianfuh.ong@eli-np.ro
[b]kmhock@utm.my

Abstract. High-power lasers can produce high-energy gamma rays, charged particles, and neutrons and induce various types of nuclear reactions. In Extreme Light Infrastructure Nuclear Physics (ELI-NP), Romania, high-power lasers are entering a new realm of 10 PW peak power, capable of obtaining a focused intensity of 10^{23} Wcm^{-2}. Such an intense laser pulse will be used for studies relevant to nuclear physics, high-field physics, and quantum electrodynamics, or the combination of laser gamma experiments. Here, we describe how a laser is used to drive high-energy photons and accelerate electrons and protons. These particles can be used for secondary interactions in nuclear physics. Laser-driven nuclear physics can be a source of nuclear isomers for applications in medicine and astrophysics.

1. Introduction

The laser-nuclear interaction can be categorized into direct and indirect interactions similar to inertial confinement fusion science. In the direct interaction, the strong laser field interacts directly with the nucleus [1][2], changes the state of decay products [3], or induces electromagnetic deexcitation of the nucleus [4][5]. In the indirect interaction, the laser first produces high energy photons, electrons, or ion beams; then these beams interact with the nucleus to make a nuclear state transition. The laser-driven nuclear physics at high-power laser facilities relies on the indirect method [6].

The high-energy photon relevant to nuclear interaction, in particular the γ–ray, can be produced by various methods. One of them is the bremsstrahlung process, in which the gamma-ray is produced when a high-energy electron bunch passes through a high-Z target. The electron bunch can be obtained from plasma-based accelerators instead of kilometer-sized conventional accelerators. The plasma-based or more specifically the Laser Wakefield Accelerator (LWFA) [7][8] is capable of sustaining many orders of magnitude higher accelerating gradients compared to the conventional accelerator, making it an attractive method as a table-top electron accelerator [9–11].

On the other hand, the ion beam can be produced via Target Normal Sheath Acceleration (TNSA) [12–14], in which a laser pulse is shot onto a solid target and the ion is accelerated by the charge separation field created by the electron at the target rear surface. The accelerated ion beam with an average energy of a few MeV can be obtained [15–18].

These high-energy beams are used for the production of nuclear isomers through (p,n), (γ,γ′) or (γ,n) channels, for medical applications and astrophysics studies. In the following sections, we describe the physics of LWFA and TNSA. Examples of nuclear isomer production for medical applications and topics relevant to astrophysics are discussed.

2. Laser wakefield acceleration

In LWFA experiments, a laser pulse is shot into the underdense plasma. The laser pulse propagates and generates a strong wake wave, which allows electrons to surf to high energy within a centimeter-scale distance [7][8]. There are several regimes of LWFA that are characterized according to the laser power, energy, and intensities [19]. In particular, the bubble regime is where electrons are pushed aside by the laser ponderomotive force and an ion cavity is formed as shown in Fig. 1. This ion cavity or bubble can sustain a few GVm^{-1} longitudinal electric fields due to the plasma's already broken state. The requirement for the bubble regime is that the laser power exceeds the critical power for relativistic self-focusing to provide stable laser beam propagation along the plasma channel [20]. The threshold power reads:

$$P > P_{bubble} = (\tau_L[fs]/\lambda_L[\mu m])^2 30[GW] \quad (1)$$

A typical short pulse laser with $\tau_L = 25$ fs and central wavelength $\lambda_L = 0.8$ μm gives $P_{bubble} > 30$TW. The force of the ion cavity is required to approximately balance the laser ponderomotive force with a match condition where $w0 \sim c\tau_L \sim \sqrt{a_0}$, where w_0 is the laser Gaussian beam waist radius, τ_L is the laser pulse duration, c is the speed of light in vacuum, and a_0 is the normalized laser field amplitude. Figure 1 shows the results of the three-dimensional (3D) particle-in-cell (PIC) simulation with 100 TW, 25 fs laser pulse propagating along 3 mm helium plasma. This simulation was performed using the picongpu [21] code with 16 V100 Nvidia GPUs. A nears-pherical bubble covering the laser pulse can be seen and is followed by plasma wakes. An electron beam of several hundreds of MeV is trapped inside the bubble after the laser has propagated for 2-3 mm in the plasma. The accelerating field along the laser propagation Ex reaches 730GVm−1 as is evident from the lineout. Figure 2 shows an electron bunch with energies up to 320 MeV and 1.4 nC of charge that can be obtained after exiting the plasma channel.

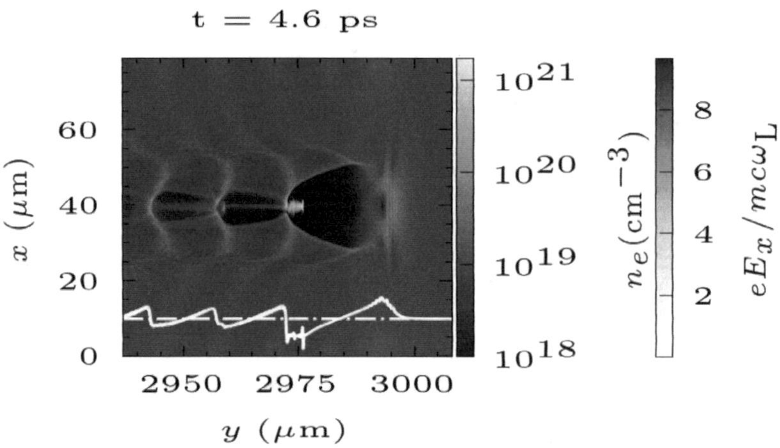

Figure 1. Simulation result for LWFA in the bubble regime with a laser of $a_0 = 2.4$. A 2D slice on the x-y plane with electron density and laser envelope after t = 4.6 ps propagation into the helium plasma. The lineout is the longitudinal electric field, E_x (normalized) along the center of the simulation. One tick corresponds to $730GVm^{-1}$. This simulation was performed by using PICONGPU [21] code with 16 V100 Nvidia GPUs.

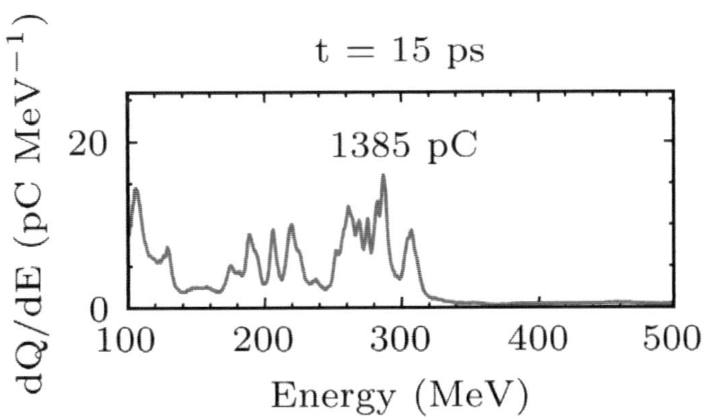

Figure 2. The energy spectrum of LWFA electron bunch exiting the helium plasma. The electron has energies up to 320 MeV. The total electron charge is 1385 pC.

3. Target normal sheath acceleration

In TNSA experiments, a laser pulse with intensity $I \geq 10^{18}Wcm^{-2}$ is shot onto a solid target of a few micrometer thicknesses. The laser prepulse with lower intensities arriving prior to the mentioned high-intensity main pulse heats and generates preplasma on the front and rear surfaces of the target. Upon the arrival of the main pulse, it interacts with the pre-plasma and generates fast electrons mainly through resonance absorption [22], vacuum heating [23], or **J** × **B** heating [24]. These fast electrons propagate across the target and exist on the rear surface, exhibiting a Maxwellian distribution. The electrons escaping form a charge separation (sheath) field of order TVm^{-1} on the rear surface, thus accelerating the proton or ion from the contamination layer. Figure 3 (a) shows the results of the two-dimensional (2D) PIC simulation in which a 1 PW laser pulse with $I = 10^{21}Wcm^{-2}$ incident on a 20 μm thick aluminum target at an oblige angle of 26^0. This simulation was performed using a PICONGPU [21] code with 8 V100 Nvidia GPUs. Electrons exiting the rear side of the target can be seen. This creates the sheath field on the rear side, as shown in Fig. 3 (b), where its magnitude is of the order of $1TVm^{-1}$. Figure 3 (c) shows the distribution of proton density that begins to accelerate forward by the sheath field. A Maxwellian distributed proton beam with cut-off energy at 27 MeV is obtained as shown in Fig. 4.

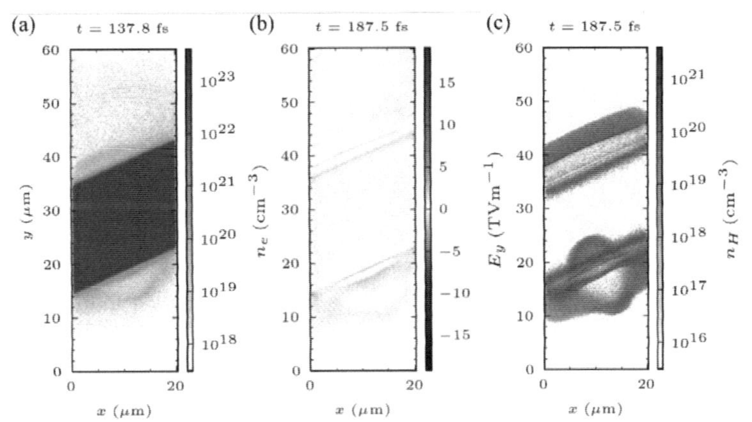

Figure 3. Simulation results for a TNSA with a laser of $a_0 = 21$. (a) The electron density distribution of 20 μm thick aluminum target after laser irradiation at an oblige angle of 26^0. The laser is linearly polarized in the x direction. (b) The longitudinal electric field, E_y at the front and rear side of the target after t = 187.5 fs of interaction. (c) The proton density distribution from the contamination layer at the front and rear surface to the target. This simulation was performed by using PICONGPU [21] code with 8 V100 Nvidia GPUs.

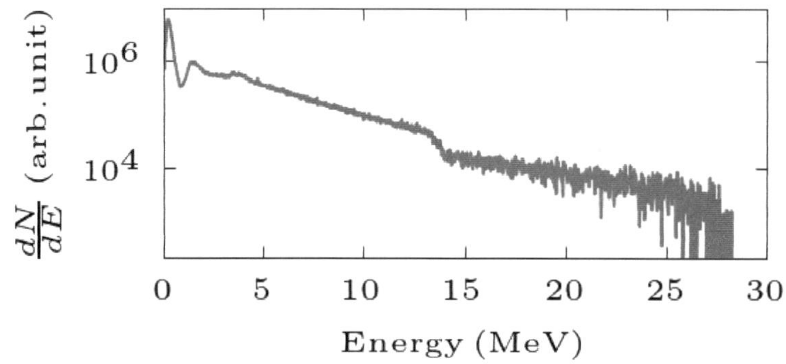

Figure 4. The proton energy spectrum.

4. Nuclear isomer production

The electron beam from the LWFA experiments will be transported through a dipole magnet to separate itself from the remainder of the laser pulse. Then, this beam is sent to a high-Z material, typically tungsten or lead for bremsstrahlung production. The proton beam from TNSA experiments will pass through a set of focusing optic. The γ-ray from bremsstrahlung or the proton beam is irradiated on nuclear target to produce nuclear isomers. The candidates of nuclear isomers include but not limited to the list in Table. 1. These nuclear isomers undergo the reaction with bremsstrahlung photons through the (γ,γ′) or (γ,n) channels. Figure 5 (a) shows the spectrum of bremsstrahlung photons where the electron beam in Fig. 2 passes through a 6 mm lead target. The photons have the energies up to 150 MeV. Electron beam transport and bremsstrahlung production are calculated using the geant4 code.

Table 1. The candidates of nuclear isomers.

Element	Level	E (keV)	Half-Life	Spin and Parity
^{139}Nd	2	231.15(5)	5.50(20) h	11/2$^-$
^{152}Eu	1	45.5998(4)	9.3116(13) h	0$^-$
^{176}Lu	1	122.845(4)	3.664(19) h	1$^-$
^{182}Hf	9	1172.88(18)	61.5(15) h	8$^-$

Figure 5(b) shows the cross sections for production of ^{182,m}Hf through the (γ, n) reaction channel as a function of incident photon energy. The cross section are calculated using the TALYS code [25]. The cross section peaks at energies between 10–30 MeV, which means that the photon required between these energies should be large to have high nuclear isomer production. The total yield for the production of ^{182,m}Hf is shown in Fig. 5(c). The production of the isomer saturates at about 80 MeV with the steepest slope around the cross-sectional peak.

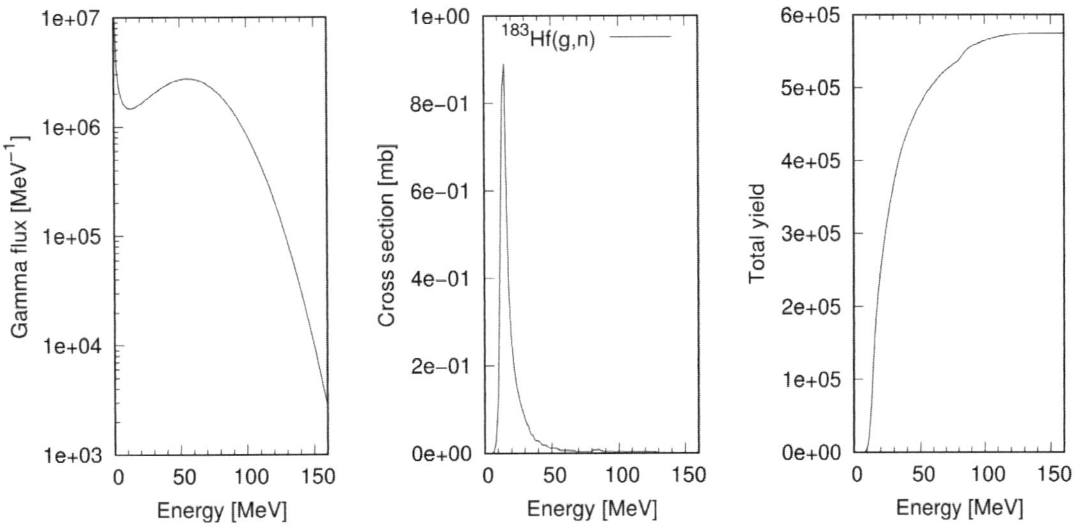

(a) (b) (c)

Figure 5. (a) The bremsstrahlung photon energy spectrum after electron beam from LWFA passes through the 6 mm lead target. The bremsstrahlung production and electron beam transport are calculated using GEANT4 code. (b) Cross sections of ^{183}Hf(γ, n)^{182,m}Hf reaction as a function of incident photon energy calculated using the TALYS code. (c) The total yield of ^{182,m}Hf production showing a plateau at about a photon energy of 100 MeV.

5. Nuclear isomer production

The electron beam from the LWFA experiments will be transported through a dipole magnet to separate itself from the remainder of the laser pulse. Then, this beam is sent to a high-Z material, typically tungsten or lead for bremsstrahlung production. The proton beam from TNSA experiments will pass through a set of focusing optic. The γ-ray from bremsstrahlung or the proton beam is irradiated on nuclear target to produce nuclear isomers. The candidates of nuclear isomers include but not limited to the list in Table. I. These nuclear isomers undergo the reaction with bremsstrahlung photons through the (γ,γ′) or (γ,n) channels. Figure 5 (a) shows the spectrum of bremsstrahlung photons where the electron beam in Fig. 2 passes through a 6 mm lead target. The photons have the energies up to 150 MeV. Electron beam transport and bremsstrahlung production are calculated using the geant4 code.

Figure 5(b) shows the cross sections for production of ^{182,m}Hf through the (γ, n) reaction channel as a function of incident photon energy. The cross section are calculated using the TALYS code [25]. The cross section peaks at energies between 10–30 MeV, which means that the photon required between these energies should be large to have high nuclear isomer production. The total yield for the production of ^{182,m}Hf is shown in Fig. 5(c). The production of the isomer saturates at about 80 MeV with the steepest slope around the cross-sectional peak.

6. Conclusions

Herewith, we have shown that the production of radioisotopes in their isomeric states through laser-driven photon generation can be explored experimentally. Another possibility to explore is the photoexcitation of isomers from excited levels by another γ- ray source. This could provide insight into the reaction rate itself, which is beyond the capability of current technology. This can improve our understanding of the photonuclear reactions that take place in massive stars and supernovae.

Another possible extension of the current project would be to employ the proton beam from the TNSA experiments for ^{99,m}Tc production through (p,2n) reaction on ^{100}Mo. The competition yield between the reaction channels (p,2n) and (γ,n) can also be tested. The radioisotopes Technetium-99m is the major ingredient for nuclear medicine imaging technique in Single Photon Emission Computed Tomography (SPECT). The quest of new production methods arises due to the closure of nuclear reactor that provides ^{99}Mo which decay to ^{99,m}Tc via beta-decay.

In summary, laser-driven plasma-based accelerators offer table-top high-energy electron, proton, and γ-ray beams that open the possibilities for studying various nuclear reactions relevant to astrophysics and nuclear medicine.

7. References

[1] Arad B, Eliezer S and Paiss Y 1979 Nuclear "anti-stokes" transitions induced by laser radiation, *Phys. Lett. A* **74** 395.
[2] Becker W, Schlicher R and Scully M 1984 Laser-induced nuclear anti-stokes transitions revisited, *Phys. Lett. A* **106** 441.
[3] Becker W, Schlicher R and Scully M 1984 Forbidden nuclear β-decay in an intense plane-wave field *Phys. Lett. A* **426** 125.
[4] Baldwin G C and Wender S A 1982 Effect of laser radiation on the decay of 235mU, *Phys. Rev. Lett.* **48** 1461.
[5] Kalman P and Lovas 1987 I Laser-induced nuclear transitions, Journal of Physics G: Nucl. Phys. **13** 495.
[6] Tanaka K A, Spohr K M, Balabanski D L, Balascuta S, Capponi L, Cernaianu M O, Cuciuc M, Cucoanes A, Dancus I, Dhal A, Diaconescu B, Doria D, Ghenuche P, Ghita D G, Kisyov S, Nastasa V, Ong J F, Rotaru F, Sangwan D, Soderstrom P A, Stutman D, Suliman G, Tesileanu O, Tudor L, Tsoneva N, Ur C A, Ursescu D and Zamfir N V 2020 Current status and highlights of the eli-np research program, Matter and Radiat. at Extremes **5** 024402.
[7] Tajima T and Dawson J M 2009 Laser electron accelerator, Phys. Rev. Lett. 43, 267 (1979).
[8] E. Esarey, C. B. Schroeder, and W. P. Leemans, Physics of laser-driven plasma-based electron accelerators *Rev. Mod. Phys.* **81** 1229.
[9] Faure J, Glinec Y, Pukhov A, Kiselev S, Gordienko S, Lefebvre E, Rousseau J P, Burgy F and Malka V 2004A laser–plasma accelerator producing monoenergetic electron beams *Nature* **431** 541.
[10] Geddes C, Toth C, Van Tilborg J, Esarey E, Schroeder C, Bruhwiler D, Nieter C, Cary J and Leemans W 2004 High-quality electron beams from a laser wakefield accelerator using plasma-channel guiding *Nature* **431** 53.
[11] Mangles S P, Murphy C, Najmudin Z, Thomas A G R, Collier J, Dangor A E, Divall E, Foster P, Gallacher J, Hooker C *et al.* Monoenergetic beams of relativistic electrons from intense laser–plasma interactions *Nature* **431** 535 (2004).
[12] Passoni M and Lontano M 2008 Theory of light-ion acceleration driven by a strong charge separation *Phys. Rev. Lett.* **101** 115001.
[13] Hegelich B M, Albright B, Cobble J, Flippo K, Letzring S, Paffett M, Ruhl H, Schreiber J, Schulze R and Fernandez J 2006 Laser acceleration of quasi-monoenergetic More references mev ion beams *Nature* **439** 441.
[14] Poole P, Obst L, Cochran G, Metzkes J, Schlenvoigt H, Prencipe I, Kluge T, Cowan T, Schramm U, Schumacher D *et al.* 2018 Laser-driven ion acceleration via target normal sheath acceleration in the relativistic transparency regime New J. of Phys. **20** 013019.
[15] Zepf M, Clark E, Beg F, Clarke R, Dangor A, Gopal A, Krushelnick K, Norreys P, Tatarakis M, Wagner U *et al.* 2003 Proton acceleration from high-intensity laser interactions with thin foil targets, Phys. Rev. Lett. **90** 06480.
[16] Clark E, Krushelnick K, Zepf M, Beg F, Tatarakis M, Machacek A, Santala M, Watts I, Norreys P and Dangor A 2000 Energetic heavy-ion and proton generation from ultraintense laser-plasma interactions with solids Phys. Rev. Lett. **85** 1654.
[17] Clark E, Krushelnick K, Davies J, Zepf M, Tatarakis M, Beg F, Machacek A, Norreys P, Santala

M, Watts I, *et al.* 2000 Measurements of energetic proton transport through magnetized plasma from intense laser interactions with solids Phys. Rev. Lett. **84** 670.

[18] Fuchs J, Antici P, d'Humieres E, Lefebvre E, Borghesi M, Brambrink E, Cecchetti C, Kaluza M, Malka V, Manclossi M. *et al.* 2006Laser-driven proton scaling laws and new paths towards energy increase *Nature Phys*. **2**, 48.

[19] Lu W, Tzoufras M, Joshi C, Tsung F S, Mori W B, Vieira J, Fonseca R A and Silva L O 2007 Generating multi-gev electron bunches using single stage laser wakefield acceleration in a 3d nonlinear regime *Phys. Rev. ST Accel. Beams* **10** 061301.

[20] Sun G, Ott E, Lee Y C and Guzdar P 1987 Self-focusing of short intense pulses in plasmas, *The Phys. of Fluids* **30** 526 https://aip.scitation.org/doi/pdf/10.1063/1.866349.

[21] Burau H, Widera R, Honig W, Juckeland G, Debus A, Kluge T, Schramm U, Cowan T E, Sauerbrey R and Bussmann M 2010 Picongpu: A fully relativistic particle-incell code for a gpu cluster *IEEE Trans. on Plasma Sci.* **38** 2831.

[22] Gibbon P 2005 Short pulse laser interactions with matter: an introduction (World Scientific Publishing Company).

[23] Brunel F 1987 Not-so-resonant, resonant absorption *Phys. Rev. Lett.* **59** 52.

[24] Kruer W L and Estabrook K 1985 J×b heating by very intense laser light *The Phys. of Fluids* **28** 430.

[25] Koning A and Rochman D 2012 Modern nuclear data evaluation with the talys code system *Nucl. Data Sheets: Special Issue on Nuclr React. Data* **113** 2841 .

Acknowledgments

J. F. Ong thanks O. Tesileanu and A. C. Berceanu for their interests and supports to this work. This work was supported by Extreme Light Infrastructure Nuclear Physics (ELI-NP) Phase II, a project co-financed by the Romanian Government and the European Union through the European Regional Development Fund - the Competitiveness Operational Programme (1/07.07.2016, COP, ID 1334), and the Proiect Nucleu PN 19060105 faza 39/202.

Study the closed orbit correction of a compact hadron driver for cancer therapy

Leo Kwee Wah[1,a)], Ken Takayama[2], Toshikazu Adachi[2], Tadamichi Kawakubo[2] and Tanuja Dixit[3]
[1]Malaysian Nuclear Agency, Bangi 43000 Kajang. Selangor, Malaysia.
[2]High Energy Accelerator Research Organization (KEK), Tsukuba, Ibaraki, Japan.
[3]Society for Applied Microwave Electronics Engineering & Research (SAMEER), Mumbai, India

[a]leo77@nuclearmalaysia.gov.my

Abstract. A design of a compact hadron driver for future cancer therapies based on the induction synchrotron concept is given. In order to realize a slow extraction technique in a fast cycling synchrotron, which allows the energy sweep beam scanning, the zero momentum-dispersion *D(s)* region and high flat *D(s)* region are necessary. The present design meets both requirements. The lattice has the two-fold symmetry with a circumference of 52.8 m, 2 m-long dispersion-free straight section, and 3 m-long large flat dispersion straight section. Assuming a 1.5 T bending magnet, the ring can deliver heavy ions of 200 MeV/au at 10 Hz. Details of the lattice parameters and orbit correction approach and method are discussed.

1. Introduction

A lattice design of a compact hadron driver for future cancer therapies is given as shown in Figure 1 has been designed [1]. Whereby the lattice has the two-fold symmetry with a circumference of 52.8 m, 2 m long dispersion free straight section, and 3 m long large flat dispersion straight section [1] [2]. The lattice parameter is design based on the linear term of motion equation. Hence, the individual common components will be treated as 3 X 3 matrices as the following equations (1)(2)(3) and (4).

For drift space,

$$= \begin{bmatrix} 1 & L & 0 \\ 0 & 1 & 0 \\ 0 & 0 & 1 \end{bmatrix} \dots\dots\dots \qquad (1)$$

For focusing magnet,

$$= \begin{bmatrix} cos[\sqrt{k}L] & 1/\sqrt{k}sin[\sqrt{k}L] & 0 \\ -\sqrt{k}sin[\sqrt{k}L] & cos[\sqrt{k}L] & 0 \\ 0 & 0 & 1 \end{bmatrix} \dots\dots\dots. \qquad (2)$$

For defocusing magnet,

$$= \begin{bmatrix} cosh\left[\sqrt{|k|}L\right] & 1/\sqrt{|k|}sinh\left[\sqrt{|k|}L\right] & 0 \\ \sqrt{|k|}sinh\left[\sqrt{|k|}L\right] & cosh\left[\sqrt{|k|}L\right] & 0 \\ 0 & 0 & 1 \end{bmatrix} \ldots\ldots\ldots \quad (3)$$

For bending magnet,

$$= \begin{bmatrix} cos[\theta] & \rho sin[\theta] & \rho(1 - cos[\theta]) \\ -sin[\theta] & cos[\theta] & sin[\theta] \\ 0 & 0 & 1 \end{bmatrix} \ldots\ldots\ldots \quad (4)$$

where:

$k = k - value of the related magnets$
$L = Length of the related magnets$

For perturbation term, which can be treated as nonlinear term and it is an essential term must be evaluated, so that the desired beam can be accelerated inside the accelerator ring with the desired lattice. Therefore, the magnet alignment error as one of perturbation term of the particles must be considered and will be discussed at the following section.

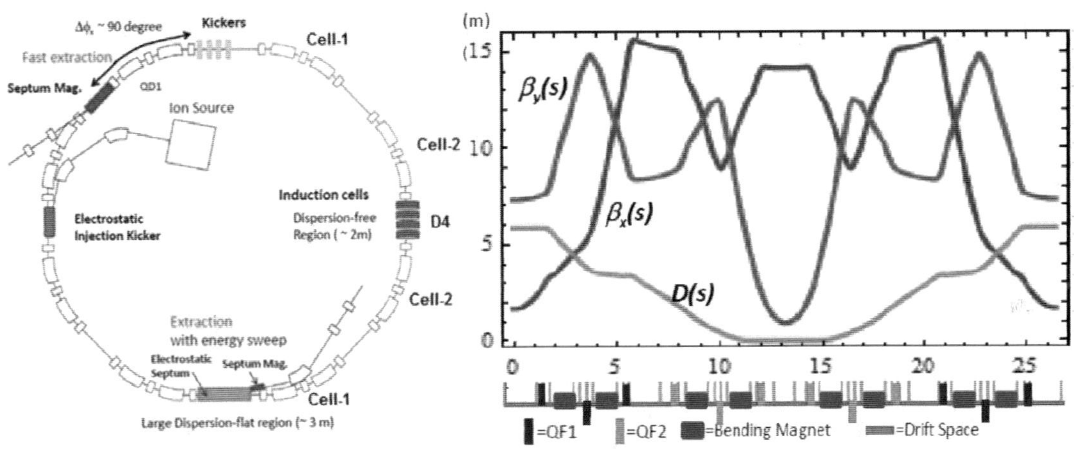

Figure 1. Outline of the ring and its lattice [1]

2. Closed orbit distortion

In general, motion of particle is a synchrotron is dominated by the betatron oscillation equation as shown in left-hand side of Equations (5) and (6) [5][6][7][8]. Without any perturbations, the beam trajectory is ideal with the movement at the orbit according to the lattice designed. If the perturbations of beam occurred in the beam transportation even in the absence of beam momentum deviation, the perturbation terms (5) and (6) are dedicated at the right-hand side as shown in Equations (7) and (8) [5][7][8].

For horizontal direction,

$$\frac{d^2x}{ds^2} + K_x(s)X = F(s') \ldots. \quad (5)$$

For vertical direction,

$$\frac{d^2y}{ds^2} + K_y(s)Y = F(s') \ldots. \quad (6)$$

After some mathematic evaluation, the closed orbit distortion (COD) deviation of a ring can be derived as (5), (6),(7) and (8):

For horizontal direction,

$$\Delta x(s) = \frac{\sqrt{\beta_x(s)}}{2sin(Q_x/2)} \int_s^{S+c} F(s')\sqrt{\beta_x(s')}\, cos\left[\frac{Q_x}{2} - |\Psi_x(s) - \Psi_x(s')|\right] ds' \ldots (7)$$

For vertical direction,

$$\Delta y(s) = \frac{\sqrt{\beta_y(s)}}{2sin(Q_y/2)} \int_s^{S+c} F(s')\sqrt{\beta_y(s')}\, cos\left[\frac{Q_y}{2} - |\Psi_y(s) - \Psi_y(s')|\right] ds' \ldots (8)$$

whereby,

F(s') is the perturbing term
β(s), β(s') and is the beta function and perturbing beta function at position s'
Ψ(s), Ψ(s') is the phase advance and perturbing phase advance at position s'
$Q_{x,y}$ is the horizontal and vertical phase advance per revolution

Typically, the perturbation terms are the caused by the miss-alignment and field error of the dipole, and quadrupole magnets. In general. The error is caused by the n-element. Therefore, the integral of Equations (7) and (8) could be derived as a sum over the n-elements in position j. Thus, the COD at the i-position can be derived as [5]:

For horizontal direction,

$$\Delta x_i = \frac{\sqrt{\beta_i}}{2sin(Q_x/2)} \sum_{j=i+1}^{i+n} F_j\sqrt{\beta_j}\, cos\left[\frac{Q_x}{2} - |\Psi_i - \Psi_j|\right] \ldots\ldots \quad (9)$$

For vertical direction,

$$\Delta y_i = \frac{\sqrt{\beta_i}}{2sin(Q_y/2)} \sum_{j=i+1}^{i+n} F_j\sqrt{\beta_j}\, cos\left[\frac{Q_y}{2} - |\Psi_i - \Psi_j|\right] \ldots\ldots \quad (10)$$

To correct both horizontal and the vertical COD, two different groups of horizontal and vertical monitors are required to measure the orbit kick. Meanwhile, the correction magnets are required to correct or minimize the orbit distortion. Such of correction algorithm for individual horizontal and vertical COD correction is discussed in next section.

3. Closed orbit distortion correction

Basically, whole set of COD correction system required combination of monitor and corrector devices. As Beam Position Monitors (BPM) will be utilized to measure the perturbation kicks in horizontal and vertical directions respectively. Meanwhile, the location of the BPM along the ring must be considered as follow (5), (6),(7), and (8):

Criteria of BPM position as:

i) To increase the signal sensitivity of the BPM at horizontal direction; Large βx is required
ii) To increase the signal sensitivity of the BPM at vertical direction; Large βy is required
iii) To avoid the shift due to the momentum deviation, free dispersion should be considered

By considering the closed orbit positions (This distortion is caused by lattice components such as dipole and quadrupole field and alignment errors) are observed at m points as the form of vector components b_i (i=1, 2, 3...... m) as:

$$b = \begin{bmatrix} b_1 \\ \vdots \\ b_m \end{bmatrix} \dots\dots \quad (11)$$

For the correction devise, the corrector magnets are required in horizontal and vertical directions. By introduction of the correction magnet as l. The correcting kicks are representing by a vector θ with components θ_j, (j=1, 2, 3 ......n),

$$\theta = \begin{bmatrix} \theta_1 \\ \vdots \\ \theta_n \end{bmatrix} \dots\dots \quad (12)$$

The orbit distortion at the i-th observation point due to the kick of the j-th magnet is known as shown in Equation (13). Those magnets are described by a matrix of A (m x n).

$$a_i = \frac{\sqrt{\beta_i}\sqrt{\beta_j}}{2sin(Q_x/2)} cos\left[\frac{Q_y}{2} - |\Psi_i - \Psi_j|\right] \dots\dots \quad (13)$$

The action of θ on the closed orbit is A_x. The correction is aim to reducing the peak-to-peak distortion of the residual orbit.

$$R = b + A\theta \dots\dots \quad (14)$$

By using the Least-Square-Fitting method into (14),

$$\chi^2 = \sum_{i=1}^{M} \quad (15)$$

with M=Number of observation points which caused by the kick of the correctors (X or Y).

It is minimum if the derivatives with respect to each component components of θ_j is zero.

$$\frac{\partial \chi^2}{\partial \theta_j} = 0 \dots\dots \quad (16)$$

$$(\textstyle\sum_{i=1}^{M} r_i\, a_{ij} = \sum_{i=1}^{M} b_i a_{ij} + \quad (17)$$

(Equation (17) is for vertical direction).

By setting R=0, from (17), (vertical direction),

$$\sum_{i=1}^{M} b_i a_{ij} = - \quad (18)$$

For the vertical direction, b and a is well-defined from the magnet error and lattice parameters. By solving the Equation (18), the kick θ can be identified individually. Meanwhile at the horizontal direction, a shift might be contributed to the distortion due to momentum deviation of the particles. Therefore, an offset term is required to compensate the shift as shown in Equation (19).

$$x_{eq} = \frac{\Delta p}{p} D(s) \dots\dots \quad (19)$$

From Equations (17) and (19) and R=0, (Horizontal direction):

$$\sum_{i=1}^{M} \quad (20)$$

By solving the Equation (18) with additional distortion as Equation (19), the kick θ as shown in Equation (20) can be identified individually.

4. Simulation approach

In this simulation, the errors of the 24 quadrupole magnets as shown in Figure 2 with the relative location as shown in Table 1 due to the alignment or field errors are generated randomly by using the Gaussian distribution.

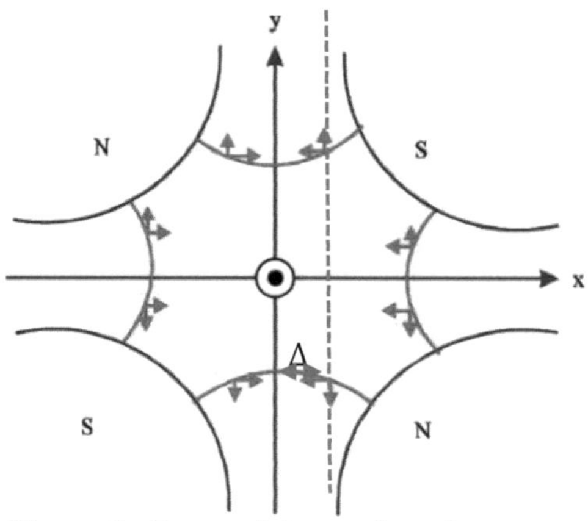

Figure 2. Errors of the quadrupole magnet.

From the Figure 2, the error field can be derived as:

$$F(s) = \Delta k(s) \dots\dots (21)$$

By introducing Equations (21) into (7) and (8):

For horizontal direction,

$$\Delta x_i = \frac{\sqrt{\beta_i} l}{2 sin(Q_x/2)} \sum_{j=i+1}^{i+n} \Delta k_{x,j}(s) \sqrt{\beta_j} \, cos\left[\frac{Q_x}{2} - |\Psi_i - \Psi_j|\right] \dots\dots \quad (22)$$

For vertical direction,

$$\Delta y_i = \frac{\sqrt{\beta_i} l}{2sin(Q_y/2)} \sum_{j=i+1}^{i+n} \Delta k_{y,j}(s) \sqrt{\beta_j} \, cos\left[\frac{Q_y}{2} - |\Psi_i - \Psi_j|\right] \ldots\ldots \quad (23)$$

whereby,

Δ is generated randomly by using Gaussian distribution approach with sigma as 10μ

l is length of the quadrupole. In addition, 24 Beam Position Monitor (BPM) are located at the position as shown in Figure 3 and Table 1.

Instead of identifying the COD at the specific location, the COD along the ring could be identified by modification of Equations (22) and (23). The modifications are shown in Equations (24) and (25):

For horizontal direction,

$$\Delta x(s) = \frac{\sqrt{\beta_x(s)} l}{2sin(Q_x/2)} \int_s^{S+c} \Delta k_x(s') \sqrt{\beta_x(s')} \, cos\left[\frac{Q_x}{2} - |\Psi_x(s) - \Psi_x(s')|\right] ds' \ldots (24)$$

For vertical direction,

$$\Delta y(s) = \frac{\sqrt{\beta_y(s)} l}{2sin(Q_y/2)} \int_s^{S+c} \Delta k_y(s') \sqrt{\beta_y(s')} \, cos\left[\frac{Q_y}{2} - |\Psi_y(s) - \Psi_y(s')|\right] ds' \ldots (25)$$

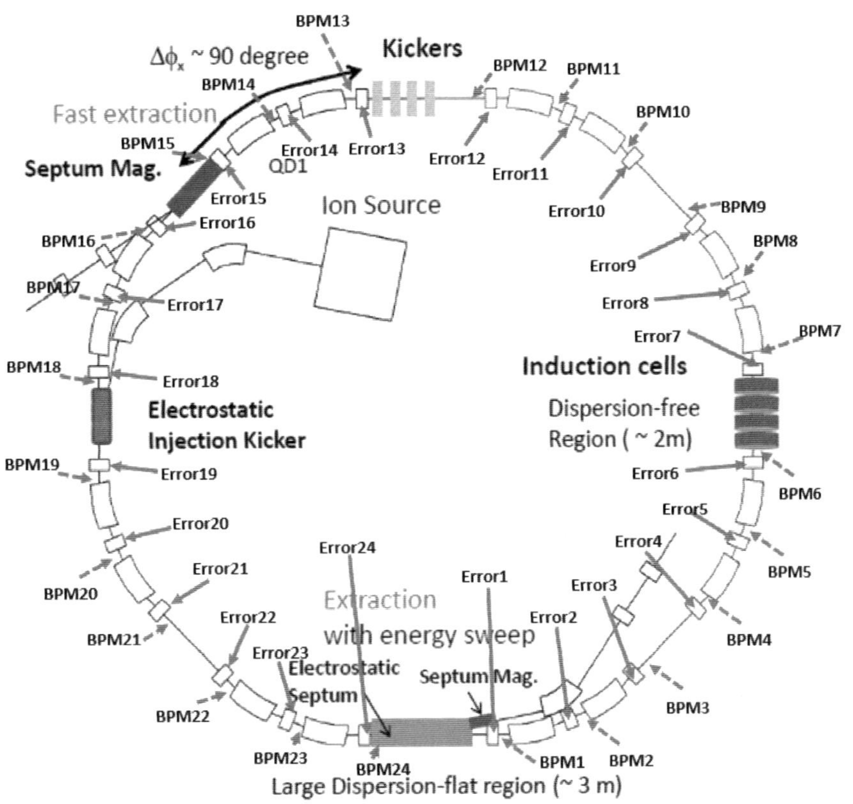

Figure 3. Hadron driver system.

Table 1. Location of the 24 errors and 24 BPM.

Error [j]	S_j[m]	BPM[i]	S_i[m]	Erros [j]	S_j[m]	BPM[i]	S_i[m]
1	1.65	1	1.95	13	28.0548	13	28.3548
2	3.6756	2	3.9756	14	30.0804	14	30.3804
3	5.7012	3	6.0012	15	32.1060	15	32.406
4	8.0012	4	8.3012	16	34.4060	16	34.706
5	10.0268	5	10.3268	17	36.4316	17	36.7316
6	12.0524	6	12.3524	18	38.4572	18	38.7572
7	14.3524	7	14.6524	19	40.7572	19	41.0572
8	16.378	8	16.678	20	42.7828	20	43.0828
9	18.4036	9	18.7036	21	44.8084	21	45.1084
10	20.7036	10	21.0036	22	47.1084	22	47.4084
11	22.7292	11	23.0292	23	49.1340	23	49.434
12	24.7548	12	25.0548	24	51.1596	24	51.4596

From the assumption as above, the number and strength of correctors to minimize the COD could be identified with Equations (18) and (20) as described in next section with two cases.

5. Result and discussion

In general, the approach based on the Equations (22) and (23) are used to evaluate the number of correction magnets and its strength and to determine the best combination to correct the COD as follow.

5.1 Case A

In this lattice, the eight corrector magnets with the different parameters as θ(s) are inserted as shown in Figure 4 and Table 2.

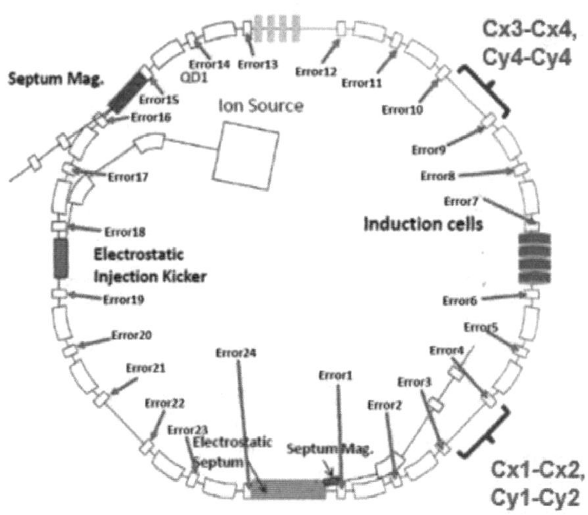

Figure 4. Case A: Eight corrector magnets (x as horizontal and y as vertical) are inserted into the ring

Table 2. Location of 4 horizontal and 4 vertical corrector magnets.

CX[i]	Si[m]	Cy[i]	Si[m]
1(H)	6.3012	1(V)	6.4512
2(H)	6.6012	2(V)	6.7512
3(H)	19.0036	3(V)	19.1536
4(H)	19.3036	4(V)	19.4536

By considering the corrector magnets configuration as shown in "Case A", the required θ(s) and the results are shown in Table 3, Figure 5 and Figure 6.

Table 3. Parameter of 4 horizontal and 4 vertical corrector magnets.

C_X[i]	Θ_x[μrad]	C_y[i]	Θ_y[μrad]
1(H)	20.227	1(V)	9.85226
2(H)	- 23.5035	2(V)	- 3.13487
3(H)	- 52.8423	3(V)	70.3711
4(H)	53.9293	4(V)	- 73.8987

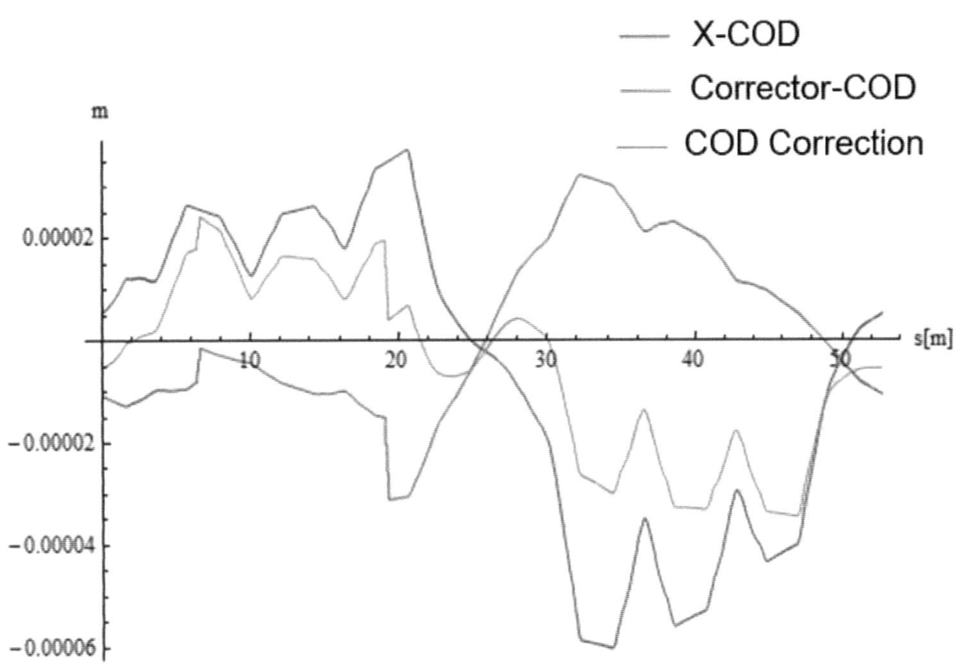

Figure 5. Horizontal: COD, COD corrector and COD correction of case A.

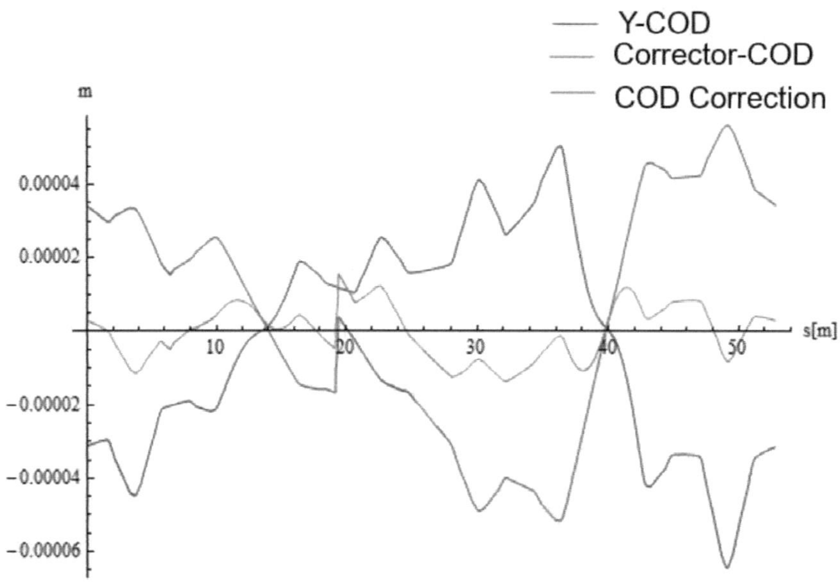

Figure 6. Vertical: COD, COD corrector and COD correction of case A.

5.2 Case B

In this lattice, the sixteenth corrector magnets with the different parameters as θ(s) are inserted as shown in Figure 7 and Table 4.

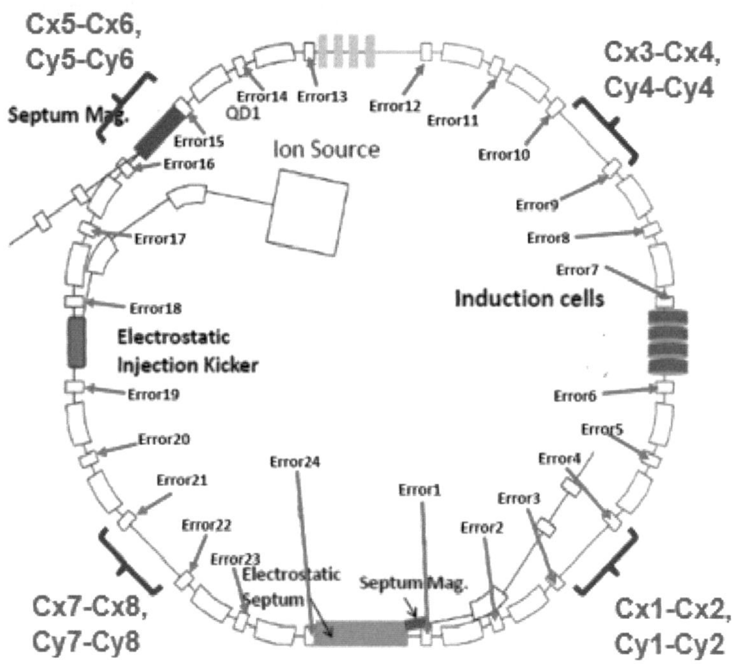

Figure 7. Case B: Eight corrector magnets (x as horizontal and y as vertical) are inserted into the ring

Table 4. Location of 4 horizontal and 4 vertical corrector magnets.

C_x[i]	S_i[m]	C_y[i]	S_i[m]
1(H)	6.3012	1(V)	6.4512
2(H)	6.6012	2(V)	6.7512
3(H)	19.0036	3(V)	19.1536
4(H)	19.3036	4(V)	19.4536
5(H)	32.706	5(V)	32.856
6(H)	33.0084	6(V)	33.156
7(H)	45.4048	7(V)	45.5584
8(H)	45.7048	8(V)	45.8584

By considering the corrector magnets configuration as shown in "Case B", the required θ(s) and the results are shown in Table 5, Figure 8 and Figure 9.

Table 5. Parameter of 8 horizontal and 8 vertical corrector magnets.

C_x[i]	Θ_x[μrad]	C_x[i]	Θ_x[μrad]	C_y[i]	Θ_y[μrad]	C_y[i]	Θ_y[μrad]
1(H)	-38.9342	5(H)	- 25.5186	1(V)	57.9945	5(V)	35.4793
2(H)	35.7962	6(H)	25.4174	2(V)	- 54.8298	6(V)	- 35.9908
3(H)	- 12.2366	7(H)	25.9818	3(V)	- 21.6066	7(V)	- 45.4047
4(H)	19.8228	8(H)	- 23.5545	4(V)	11.395	8(V)	41.7979

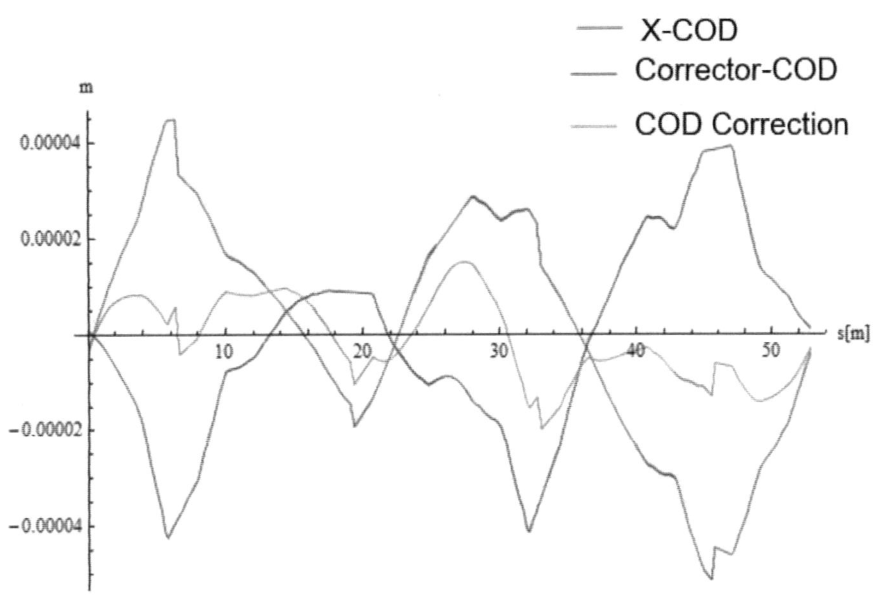

Figure 8. Horizontal: COD, COD corrector and COD correction of case B.

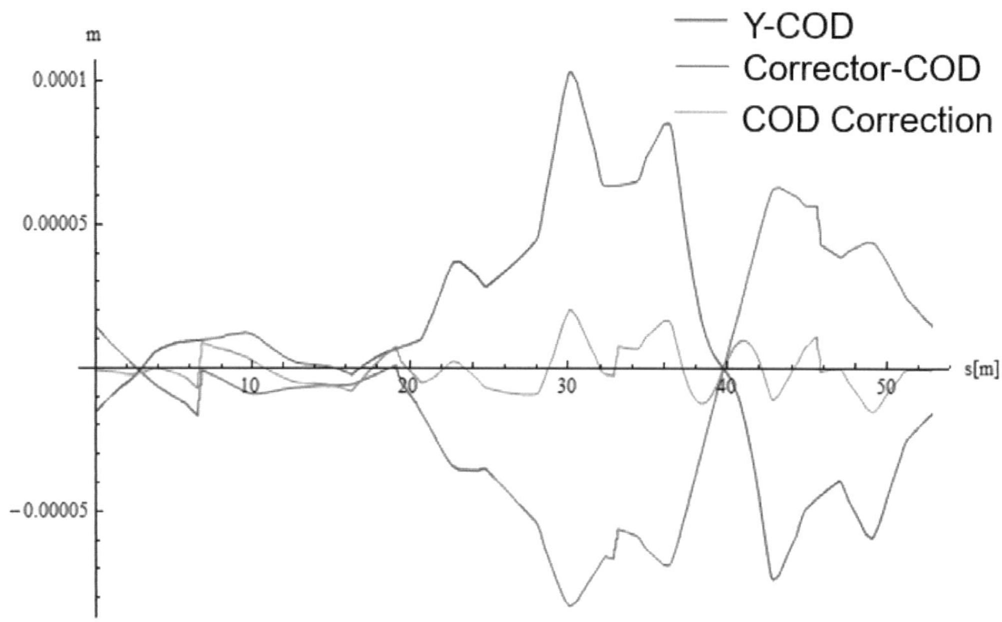

Figure 9. Vertical: COD, COD corrector and COD correction of case B.

According to Figures 5, 6, 8 and 9, we noticed that two different corrector magnet configurations able to minimize the COD caused by the miss-alignment of the dipole magnets. The COD is compensated by the COD corrector and compensation results are shown as COD correction. In ideal situation, the COD correction is required in every space along the ring. For situation, ring is occupied by the magnet components, injection and extraction and diagnostic devices. Therefore, the number of BPM and correctors is limited and must be selected within the space consideration and alignment of magnet is very critical and must be measured precisely. By using the Least-square fitting method the optimized value of correctors strength could be identified with the minimize distortion, which have been justified by case A and B,

6. Conclusion

In this paper, approach of Closed Orbit Distortion (COD) correction is introduced and evaluated. The configuration of the monitoring and corrector devices could be identified with perfectly correction. In near future, the dynamic aperture of the ring must be considered.

7. References

[1] Leo K W, Monma T, Adachi T, Kawakubo T, Dixit T and Takayama K 2016 compact hadron driver for cancer therapies using continuous energy sweep scanning *Phys. Rev. ST AB* **19**, 042802.

[2] Leo K W, Adachi T, Kawakubo T, Monma T, Dixit T and Takayama K . 2015 A compact hadron driver for cancer therapies using continuous energy sweep scanning *Proc. of the 13th Int. Conf. on Heavy Ion Accelerator Tech.* (Yokohama).

[3] Wiedemann H *Particle Accelerator Physics* 3rd ed. (Berlin Heidelberg New York: Springer) ISBN-13 978-3-540-49043-2.

[4] Adachi T 2014 *KEK-Nuclear Malaysia-Sokendai Accelerator School in Nuclear Malaysia.*

[5] Peggs S and Satogata 2017T Introduction to accelerator dynamics 1st ed (Cambridge University Press).

[6] Adachi T KEK DA Note 11-01 COD in the DA ring due to magnet disorder caused by Higashi-Nippon big disaster.

[7] Takayama K KEK DA Note 11-02 Closed orbit distortion of KEK digital accelerator ring, property of COD in a combined-function type ring
[8] Milutinovic J and Ruggiero A G 1988 Technical report. N0. 107. Closed orbit analysis for the AGS booster (Brookhaven National Laboratory).

Acknowledgments
We would like to express our gratitude to Ministry of Science, Technology and Innovation (MOSTI), Malaysia and High Energy Accelerator Research Organization (KEK), Japan for supporting the allocation to collaborate between Malaysian Nuclear Agency and KEK for this study.

X-ray imaging of an industrial microcontroller sample using dual-energy technique

Azhani Mohd Razali [1, a)], Roslan Yahya[1], Maziah Jamaludin[1]

[1] Plant Assessment Technology (PAT) Group, Industrial Technology Division, Malaysian Nuclear Agency, Bangi, 43000 Kajang, Selangor, Malaysia

azhani@nuclearmalaysia.gov.my

Abstract. In a conventional single-energy X-ray imaging technique, the information obtained from the examined object are often not sufficient for precise characterization. Hence, this study was designed to assess whether the dual-energy imaging technique would be a good alternative to the conventional method and can be used to improve the output image. An electronic microcontroller was chosen as the sample due to its complex inner structure. It was scanned using the X-ray micro computed tomography (micro-CT) scanner at two different energy levels: 60kVp and 100kVp. Dual-energy image processing method was applied to the acquired image and the results were compared to the single-energy image in term of contrast and image quality. The combined information from dual-energy scan at 60kVp and 100kVp was able to visualize the sample's internal structure (raised surface). By comparison, the single-energy imaging technique was only capable at producing shadow image. The resolution of the dual-energy image, however, was much lower due to the loss of data....

1. Introduction

X-ray imaging originated from the medical sector and since then has been extended to several industrial applications [1]. The excellent penetration ability of x-rays has made X-ray imaging a powerful imaging modality and this led to the rapid development of high-performance X-ray imaging systems [2]. Such imaging system is particularly useful for applications in the nondestructive industrial and safety inspection [3]. In recent years, more X-ray imaging technique has been explored by researchers to improve the image of complex materials and this include the dual-energy X-ray imaging technique. A conventional setting of single-energy imaging utilizes a single polychromatic X-ray spectrum (ranging from 30 kVp to 140 kVp) emitted from a single radiation source and received by a single detector [4]. On the other hand, dual-energy imaging refers to the use of two different low- and high-energy x-ray, or the use of energy-sensitive photon-counting detector. The first approach may be implemented in several ways: different spectra can be obtained by varying the parameters such as the source voltage, target materials or radiation sources from different x-ray tubes. For the second approach, the photon-counting detector is used to differentiate photons according to their energy from a single polychromatic x-ray exposure. This system is often referred to as "spectral imaging" or "color imaging".

The advantages of the dual-energy imaging rely on the additional attenuation of material after being exposed with different x-ray energies. This is possible because the attenuation of x-ray photons is energy-dependent, and each material has a specific attenuation curve [5]. Therefore, the attenuation information can be manipulated by changing the photon energy levels, allowing differentiation and classification of material-specific images. There are a growing number of interest and demand for the dual-energy research and services, however, the commercial dual-energy system is very expensive and not easily accessible to most people. It is found that the repetitions of imaging using a standard single-source imaging scanner with different energy level is the simplest way to mimic the dual-energy system, hence used in this study.

2. Materials and method

2.1. X-Ray imaging –equipment

X-ray imaging was performed using SkyScan 1172 X-ray micro-CT system stationed at the Plant Assessment Technology Group, Malaysian Nuclear Agency. Image reconstruction was executed using the NRecon software, which is the complementary software for the imaging system.

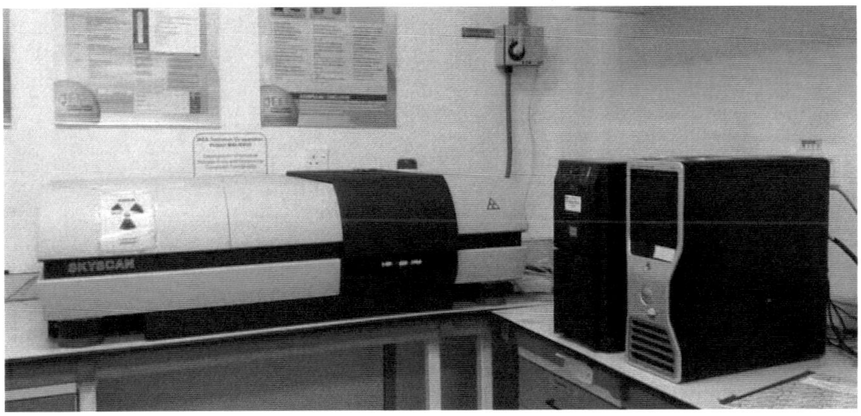

Figure 1. Micro-CT scanner.

2.2. Sample Preparation

The dimension of the microchip sample is 0.645cm x 0.520cm (approximation by vernier caliper). The microchip was placed vertically to a customized styrofoam holder to avoid the sample from moving. No other preparation was needed prior to scanning process.

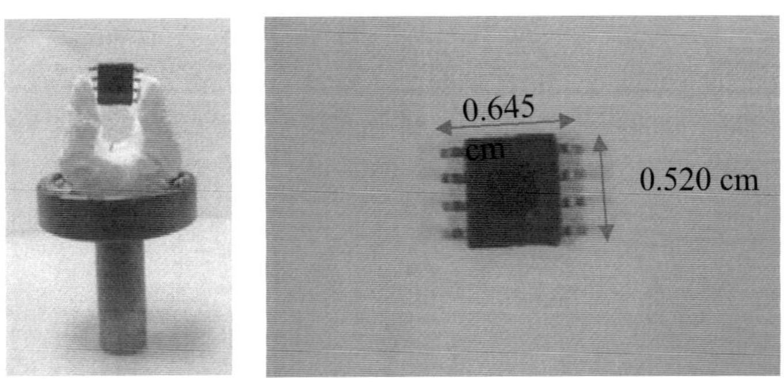

Figure 2. Microcontroller sample dimension.

IOP Conf. Series: Materials Science and Engineering 1285 (2023) 012005 doi:10.1088/1757-899X/1285/1/012005

2.3. Imaging protocol / procedure

Sample was scanned at two different energy levels, first at 100kVp (representing high energy x-ray), then at 60kVp (representing low energy x-ray). For each energy level, CT image reconstruction was performed independently. The results in form of projection image and tomography image were processed using the dual-energy image processing method. The tomography image, however, will not be considered and discussed in this paper due to the geometrical limitation of sample (flat surface) that produces high noise image.

There are variety of ways to process the low- and high-energy images. Three commonly used methods are: 1) weighted average or blended images, 2) energy subtraction, and 3) material decomposition map. Most dual-energy systems use the weighted average image method by default. It is basically a linear blend of the low- and high- energy acquisitions. In this study, weighted average and energy subtraction were selected as the dual-energy image processing method due to its straightforward method.

3. Results and discussion

Figures 3(a) and 3(b) are the 2d projection images of sample acquired by the single-energy scanner at 60kVp and 100kVp respectively. The grey contrast in the images is representing the differences in the X-ray attenuation by the microcontroller's material on different energy levels due to the variation in density of sample. Visually, both images produced similar image details except for the grey contrast information. The projection images were then processed together using two image processing methods: dual-energy averaging method and dual-energy subtraction method, resulting to images in Figure 3(c) and 3(d).

Upon further inspection on the interest area, which is the soldering part in the center region, it was observed that the image from the averaging method (Figure 4(a)) produced a very similar shadow image to single-energy projection image. On the other hand, the result from the subtraction method (Figure 4(b)) was slightly different from the single-energy image, in which it was able to visualize the raised surface of the microchip internal structure. The illustration of actual image of microcontroller was illustrated in Figure 4(c). Clearly, the subtraction method produces interesting qualitative result, but the resolution of the image was low due to the information loss during subtraction process.

The results from this study have proven that the dual-energy method was able to exploit the differences in the mass attenuation coefficients of different materials as a function of energy. It was noted that this dual-energy mimicking setup are not suitable for life or moving sample since the imaging procedure was not done simultaneously.

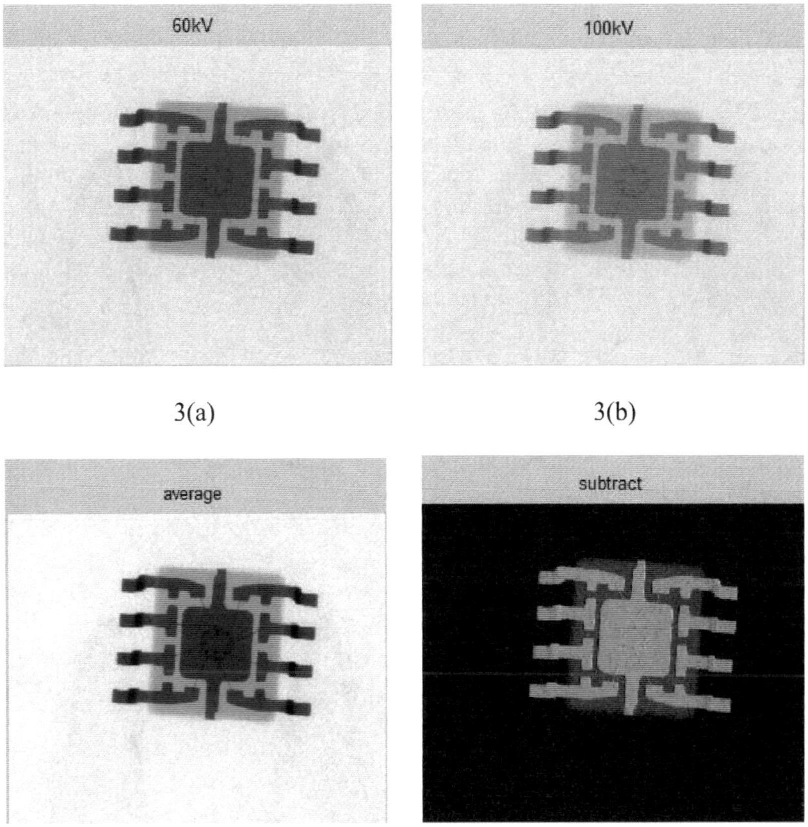

3(a) 3(b)

3(c) 3(d)

Figure 3. 2d projection image (a) at single-energy 60kVp, (b) at single-energy 100kVp, (c) using dual-energy weighted average method, and (d) using dual-energy subtraction method.

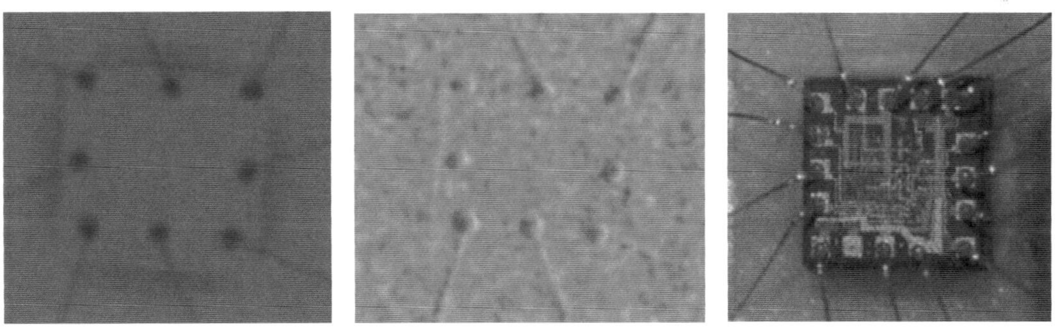

4(a) 4(b) 4(c)

Figure 4. (a) Zoomed 2D projection image of microcontroller using dual-energy weighted average method, (b) zoomed 2D projection image of microcontroller using dual-energy subtraction method, and (c) the illustration of actual image of the microcontroller.

4. Conclusion

The present study was designed to verify the capability of dual-energy technique to improve X-ray images of object under interest. Using a standard single-energy scanner, two single-energy images were acquired and combined to mimic the dual-energy system. Due to the limitation of sample geometry, only projection image was able to produce interesting and significant results. Nevertheless, the results have successfully demonstrated that the dual imaging technique is useful and capable to enhance the image details that would not be possible with single-energy technique. However, the resolution of image was quite blur and should be improved in the next study.

5. References

[1] Abdullah J *et al.* 2015 Non-destructive assaying gold jewellery using dual-energy microcomputed tomography. *Jurnal Teknologi*, **73**(3).
[2] Ou X *et al.* 2021 Recent development in X-ray imaging technology: future and challenges. research 2021 9892152.
[3] Holler M, Guizar-Sicairos M, Tsai E H *et al.* 2017 High-resolution non-destructive three-dimensional imaging of integrated circuits *Nature* **543**(7645) 402–406.
[4] Chapman H N and Nugent K A 2010 Coherent lensless X-ray imaging. nature photonics **4(**12) 833–839.
[5] Heo J H, Shin D H, Park J K, Kim D H, Lee S J, and Im S H 2018 High-performance next-generation Perovskite nanocrystal scintillator for nondestructive X-ray imaging *Adv. Mater.* **30**(40).

Acknowledgement
The present This work is registered under the Malaysian Nuclear Agency PQRD Project: NM-R&D-22-98 (no funding support). The authors would like to thank the supporting staff of Plant Assessment Technology Group and the internship student from UTM for their technical assistance.

IOP Conf. Series: Materials Science and Engineering 1285 (2023) 012006 doi:10.1088/1757-899X/1285/1/012006

The impact of policy toward R&D and innovation on nuclear technology in Malaysia

Fairuz Suzana Mohd Chachuli[1,a)] and Faridah Mohamad Idris[1]

[1]Planning & International Relations Division, Malaysian Nuclear Agency, 43000 Kajang, Selangor

[a]fairuz@nm.gov.my

Abstract. Nuclear science and technology for peaceful uses successfully progressed in various areas such as industry, medicine and healthcare, food and agriculture, natural resources, energy, and nuclear safety and security, through research and development and innovation (RDI). Realizing the importance of RDI in nuclear technology as the driving force in addressing societal challenges, most countries, including Malaysia, are developing new policies to promote R&D and innovation in nuclear technology. This paper assesses the impact of the policy on the performance of R&D and innovation in nuclear technology in Malaysia from 2011-2021 using the data envelopment analysis (DEA) and Malmquist index (MI) analysis. The study used two inputs, namely RDC & Development budget (RM Million) and Operating budget (RM Million), and three outputs, namely revenue (RM Million), no. of R&D products and no. of publications. The result shows that the performance of R&D and innovation in nuclear technology is 96.70 per cent from 2011-2021. The study shows that the growth of R&D and innovation in nuclear technology decreased by 8.3 per cent for the entire study period. The technological changes decreased at 8.3 per cent, while the efficiency changes remain. The study also found that nuclear technology growth is mainly affected by technological changes due to policy changes compare to the resources provided since efficiency changes remain.

1. Introduction

Funding for research and development (R&D) lays the foundation for rising local outputs and economic development. Arik and Ndrianasy (2018) discovered a correlation between high levels of R&D investment and high Gross Domestic Product (GDP) and given that R&D financing results in business outputs [1]. An effective research organisation will be able to demonstrate that it can convert increases in industry or government R&D funds into R&D outputs with the help of any new insights into R&D efficiency. The study of R&D funding efficiency will provide the foundation for comprehending and improving R&D activity and have a positive influence on the business community at the national level given that R&D contributes to the expansion of national economies [2]. Additionally, information on organization-level R&D efficiencies might help government decision-makers choose which organizations get more federal R&D financing.

This paper aims to create a model for estimating the technical efficiency of nuclear technology R&D activities and the R&D growth of nuclear science and technology in Nuklear Malaysia from 2011-2021. A Data Envelopment Analysis (DEA) model is developed to measure efficiency and a Malmquist Index (MI) is utilised to calculate advances in overall production. With a time horizon of around eleven years, this study offers a new framework for R&D funding effectiveness. R&D funding comes primarily from the federal government as well as private industry. In this study, Nuklear Malaysia is selected as a case study since its role as a leading R&D organization in the field of nuclear science and technology in Malaysia. This paper would significantly contribute to future strategic decision-making on the success of R&D of nuclear science and technology activities in research organizations such as Nuklear Malaysia.

The remainder of the article is structured as follows. A summary of the current research on R&D success and growth is given in Section 2. The current work's method is described in Section 3 of the article. Section 4 presents the findings and discussions. The study's summary and conclusion are presented in Section 5.

2. Literature review

Research has been done on assessing R&D performance to determine institutional effectiveness. In today's fast globalising economy, an efficient R&D organisation is a key source of competitive advantage. Thus, a significant issue for both academic and practical interests has been the evaluation of R&D performance [3]. However, due to the complex nature of these operations, it is challenging to quantify the R&D efficiency. Therefore, the DEA method is frequently employed as a nonparametric way to measure R&D efficiency [4]. Researchers are interested in R&D efficiency, and it is common to find publications on the R&D effectiveness of businesses, institutions, university research, nations, or regions.

Bae and Chang (2012) examined how open and closed innovation affected Korean companies' success [5]. They discovered that open innovation organisations had statistically higher efficiency and effectiveness than their closed counterparts. Hung and Shiu (2014) used three-stage DEA and Tobit regressions to assess the R&D effectiveness of projects in China [6]. Chun *et al.* (2015) used a two-stage DEA to analyse the effectiveness of R&D processes in Korean manufacturing companies [7]. Using the DEA and Malmquist Index, Jang (2016) calculated the cumulative change in R&D efficiency of the top R&D corporations in the technology sector [8]. Using DEA models, Khoshnevis and Teirlinck (2017) examined the effectiveness of R&D-active businesses [9]. A network data envelopment analysis was used by Qin and Du (2018) to assess the R&D performance of institutions in China's provinces [10]. Using DEA, Xiong *et al.* (2018) assessed the R&D effectiveness of Chinese research institutes [11]. Using DEA, Fairuz *et al.* (2021) investigated Malaysia's R&D performance in five renewable energy resources from 2012 to 2017 [12].

3. Methodology

3.1. Data envelopment analysis

This study employs DEA and MI analysis to evaluate the relative effectiveness of nuclear technology R&D activities and the development of nuclear science and technology in Nuklear Malaysia from 2011 to 2021. DEA is a common instrument used by academics and professionals to evaluate the performance of businesses [13-19]. DEA is a single-measure comparison approach of total performance produced by units using many inputs to produce numerous outputs which are based on linear mathematical programming [20]. According to Charnes, Cooper and Rhodes (1978), the weighted total of output to the weighted sum of input for each DMU evaluation cannot be greater than 1 [21]. The Banker-Charnes-Cooper (BCC) model and the Charnes-Cooper-Rhodes (CCR) model are two DEA models that are frequently used to calculate the level of efficiency [22-25]. Efficiency is evaluated on a scale of 0 to 1 [26]. A value of 1 indicates that a unit is reasonably efficient, whereas a value of less than 1 shows that the unit is inefficient.

This study uses the BCC-DEA model with the variable return to scale (VRS) assumption for static analysis. The MI analysis is used for dynamic analysis to assess how strategic policies have affected Nuklear Malaysia's R&D efforts in nuclear technology. Eq. (1) presents the CCR-DEA output-oriented model [26].

$Min\Phi$ is subjected to:

$$\sum_{j=1}^{n} z_i x_{ij} + s_j^- = x_0, (i = 1, \dots, m)$$
$$\sum_{j=1}^{n} z_i y_{rj} + s_j^+ = \theta_0, (r = 1, \dots, s)$$
$$z_0 \geq 0, j = 1, \dots, n$$

For VRS, add $\sum_{j=1}^{n} z_j = 1$, (1)

where n is the number of DMUs. m is the input and s is the output for each DMU_j (j = 1,2, …, n). x_{ij} (i = 1,2, …,n) is the ith input of DMU_j, and y_{rj} is the rth output of DMU_j. Slack variables, s_i^- and s_j^+, measure the excess of inputs and outputs. The efficiency value is denoted by θ_0, which is in the range of (0,1].

3.2. Malmquist Index Analysis

This study uses the Malmquist index analysis (MI) to evaluate productivity changes in nuclear technology R&D activity [24][27]. The index, which may be multiplicatively divided into an efficiency change index and a technology change index, is typically used to measure productivity change over time. Färe and Grosskopf (1984) generalised MI based on the geometric mean of two indices [28]. The productivity shift between periods t and t+1 is examined by this index [29]. Additionally, this index assesses the productivity of two distance functions by comparing their ratios [30]. The Malmquist productivity change, which consists of two components: the technical efficiency change (EFFCH) and the technological change (TECHCH), is expressed as follows in Eq. (2) [27]:

$$M_t^{t+1} = \left[\frac{D_0^t(x_0^{t+1}, y_0^{t=1})}{D_0^t(x_0^t, y_0^t)} \frac{D_0^{t+1}(x_0^{t+1}, y_0^{t+1})}{D_0^{t+1}(x_0^t, y_0^t)}\right]^{1/2}, \quad (2)$$

$$EFFCH = \frac{D_0^t(x_0^{t+1}, y_0^{t+1})}{D_0^t(x_0^t, y_0^t)}, \quad (3)$$

$$TECHCH = \left[\frac{D_0^t(x_0^{t+1}, y_0^{t+1})}{D_0^{t+1}(x_0^{t+1}, y_0^{t+1})} \frac{D_0^t(x_0^t, y_0^t)}{D_0^{t+1}(x_0^t, y_0^t)}\right]^{1/2}, \quad (4)$$

where M_t^{t+1} is the index between periods t and t+1; $D_0^t(x_0^t, y_0^t)$ and $D_0^{t+1}(x_0^t, y_0^t)$ are the distance functions between t and t+1 for inputs and outputs, respectively.

In Eq. (2), efficiency increases if M_t^{t+1}>1, remains constant if M_t^{t+1}=1 and decreases if M_t^{t+1}<1. The EFFCH component of Eq. (3) determines the catch-up impact of a specific DMU's proximity to or distance from the production frontier during periods t and t+1. The frontier shift effect, often known as TECHCH, measures the advancement or regression of DMU technology between t and t+1.

3.3. Research design

Fig. 1 depicts the conceptual framework of this study. The current study uses two variables, namely RDC & Development budget (RM Million) and operating budget (RM Million) as input for R&D activities, whilst revenue (RM Million), a number of R&D products and a number of publications are identified as an output for R&D activities. The DEA model and the Malmquist productivity index analysis are used in this study to assess the R&D performance of nuclear technology in Nuklear Malaysia, as well as to evaluate the impact of the policy interventions, on nuclear technology R&D in

Nuklear Malaysia from 2011 to 2021 using these variables. The policy interventions related to nuclear technology R&D activities considered in this study are the Malaysian Nuclear Agency Strategic Plan 2012-2020: Strategy and Action Plan [31]. Dasar Sains, Teknologi dan Inovasi Negara (DSTIN) 2013 – 2020 [32], 10th Malaysia Plan 2010-2015 (10th MP)[33] and 11th Malaysia Plan 2016-2020 (11th MP) [34]. The major incidents that affected RDI activities related to nuclear technology such as the Fukushima nuclear power plant accidents occurred in March 2011 and Covid-19 pandemic is also considered in this study.

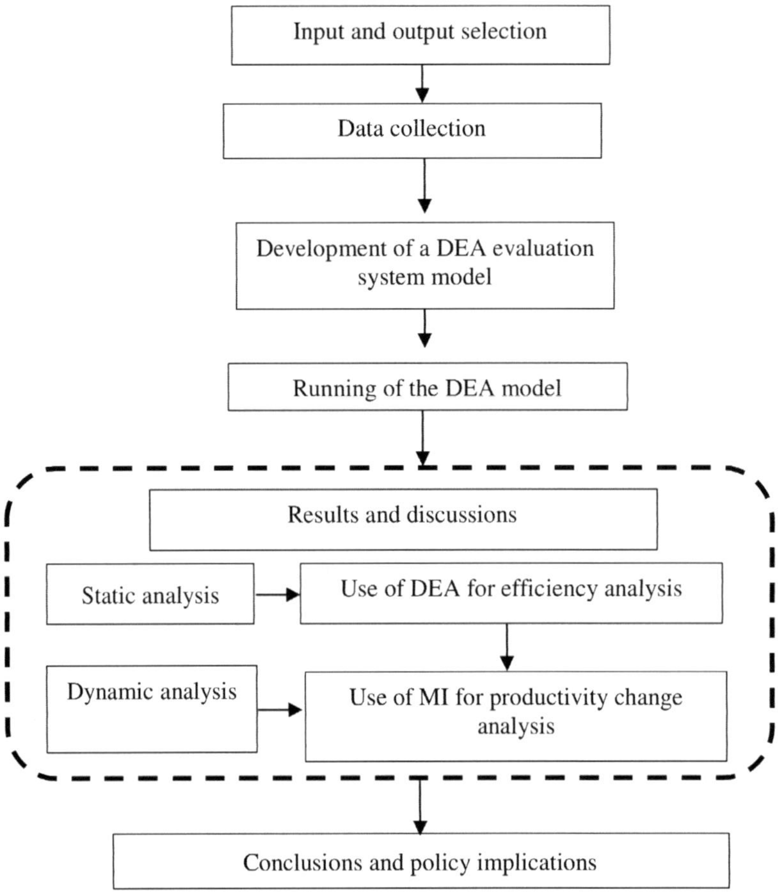

Figure 1. Proposed DEA research framework

3.4. Data collection

Table 1 shows the inputs and outputs data used collected in this study. In this study, the annual RDC & Development budget (RM Million) and Operating budget (RM Million) are selected as input from the Annual Report of the Malaysian Nuclear Agency from the year 2011-2021. Nuklear Malaysia received a total amount of RM 329.56 million for the RDC & Development budget, while RM 864.13 million for the operating budget during this period. The total government funding received for Nuklear Malaysia amounted to RM1193.69 million for 11 years. Revenue (RM Million), no. of R&D products and no. of publications are selected as an output in the study. Income generation through commercialisation activities in Nuklear Malaysia recorded a total of RM125.63 million from 2011-2021. The average total budget ratio between the funding received by Nuklear Malaysia to revenue generated through commercialisation activities in Nuklear Malaysia is around 10.52% from the period 2011-2021. It is noticed that the revenue generation for Nuklear Malaysia decreased in the year 2020 and 2021 during the Covid-19 pandemic season last two years. The number of R&D products produced by Nuklear

Malaysia shows an increasing trend over the years. From 2011-2021, the total R&D products produced from Nuklear Malaysia is 722 units including products, patents, procedures, and protocols. Another R&D output used in this study is the number of publications published by researchers of Nuclear Malaysia. The total number of publications published by Nuclear Malaysia is 5,993 units from 2011-2021, although, in the middle of the study period, the publishing trend shows a decreasing value in 2016 and 2017.

Table 1. Input and output data.

	RDC & Development budget (RM Million)	Operating budget (RM Million)	Revenue (RM Million)	No. of R&D products	No. of publications	Budget Ratio
2011	10.20	66.21	15.78	53	568	20.65%
2012	18.43	70.67	14.61	68	302	16.40%
2013	25.42	72.90	12.45	49	683	12.66%
2014	41.81	75.89	13.24	45	702	11.25%
2015	34.07	76.77	11.41	71	625	10.29%
2016	48.29	80.15	12.62	70	386	9.82%
2017	44.50	81.50	10.38	45	434	8.24%
2018	34.12	81.93	12.06	58	590	10.39%
2019	20.87	84.79	10.43	76	653	9.87%
2020	14.08	85.90	8.11	72	511	8.11%
2021	37.75	87.40	4.54	115	539	3.63%
Total	329.56	864.13	125.63	722	5993	10.52%

4. Result and discussion

4.1. Performance evaluation of nuclear technology R&D activities

Fig. 2 depicts the comparison between the budget received by Nuklear Malaysia and the revenue generated through the commercialisation activities carried out in Nuklear Malaysia from 2011-2021. The operating budget, which covers the salary of staff and daily operations, and maintenance costs is increasing every year since 2011. The increasing operating budgets are mostly due to the increasing salary of staff annually including when the staff got promoted in their grade position although the number of staff remains the same. The trend of RDC and development budget which is mainly distributed through the 5-year Malaysia Plan (MP) covers the 10th MP for 2011-2015 and 11th MP for 2016-2020 was not consistent during the study period. It seems that the RDC and development budget is high in the year 2014 during the 10th MP and the year 2016 during the 11th MP. Both years remark a high RDC and development budget received by Nuklear Malaysia when the Thorium Flagship project entitled Development of Innovative Nuclear Reactor Technology – With the Spin-Off Based on Thorium (FP0214D052 DSTIN) was implemented from 2014 to 2018. The flagship project was implemented under the Dasar Sains, Teknologi Dan Inovasi Negara (DSTIN) 2013 – 2020 due to the recommendation from the Jawatankuasa Kajian Penggunaan Torium Sebagai Bahan Asas Dalam Proses Penjanaan Tenaga Nuklear in 2013. However, the revenue generated from the commercialisation activities in Nuklear Malaysia recorded a decreasing trend over the years up to 28.76% compared to 2011 and 2021, especially during the Covid-19 pandemic phase. In addition, the decreasing trend in the revenue generated is observed due to the current efforts made by Nuklear Malaysia that have been made to ensure that research products and activities should be commercialised through technology transfer and licensing to their industrial partners.

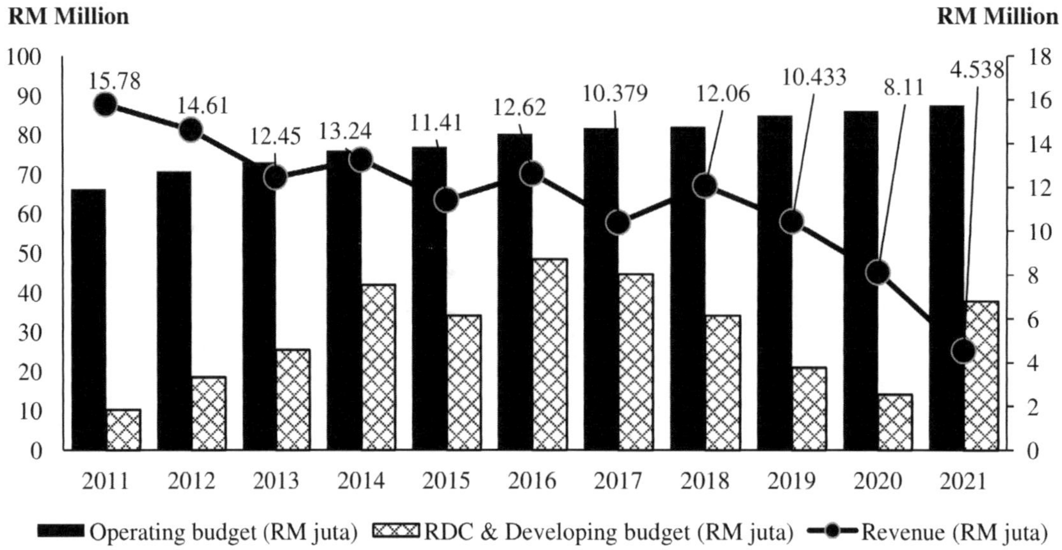

Figure 2. The budget received and revenue generated from 2011-2021.

Fig. 3 shows the trend of R&D outputs produced by Nuclear Malaysia from 2011-2020. The R&D outputs include innovation awards at national and international levels, the number of Intellectual Property Rights (IPR) generated from MOSTI grants or other grants, the number of publications, the number of products, the number of processes and procedures developed and the number of potential products for commercialisation generated from Nuklear Malaysia. The total number of R&D products produced by Nuklear Malaysia is 722 units while the number of publications published is 5,993 units from 2011-2020. Although the total publications showed a decreasing trend from 2014 to 2016, the publications activities among Nuklear Malaysia's staff are increased over the years since then. However, the total R&D products including innovation awards, IPR, products, processes, procedures, and potential products for commercialisation show an increasing trend from 2011-2020. This is showing that Nuklear Malaysia's staff are actively conducting R&D activities related to nuclear science and technology concerning the budget received for Nuklear Malaysia either from the annual operating budget or RDC & development budget.

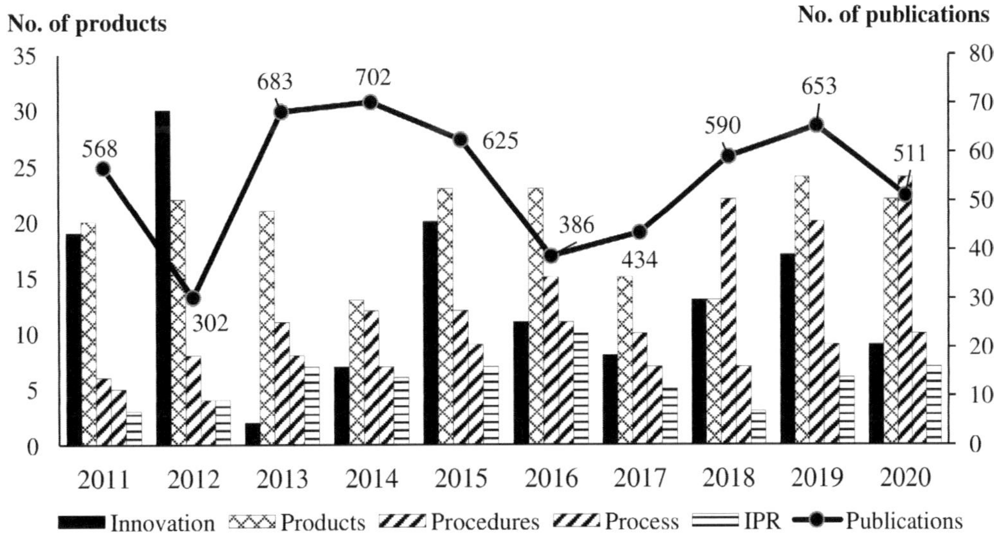

Figure 3. R&D outputs produced by the Malaysian Nuclear Agency from 2011-2020.

Fig. 4 displays a summary of the efficiency scores obtained using the DEA output-oriented model. In this study, the output-oriented model is chosen to transform an inefficient unit into an efficient one by proportionally increasing its outputs while maintaining the proportions of its inputs. The Variable Returns to Scale (VRS) is used in this study to determine whether an increase or reduction in inputs or outputs does not lead to a proportional change in the inputs or outputs, respectively. In contrast to the CRS model, the VRS model is typically used by academics as it better reflects real-world concerns. Fig.4 shows that the performance of nuclear technology R&D activities in Nuklear Malaysia is 96.70 per cent from 2011-2021. However, a decreasing trend in the R&D activities in Nuklear Malaysia is highlighted from 2016-2018 due to the decreasing number of R&D outputs produced during this period, especially total publications.

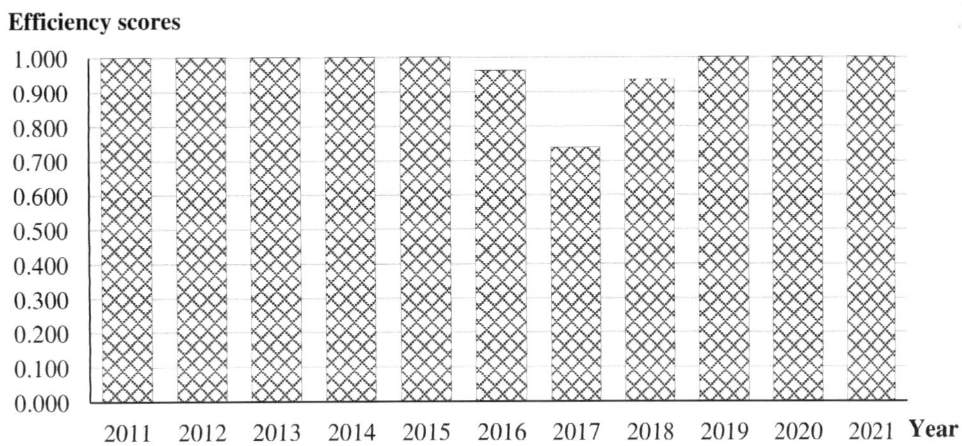

Figure 4. Summary of efficiency scores using DEA output-oriented model.

4.2. Nuclear technology R&D growth

Fig. 5 displays the trends of nuclear technology R&D's efficiency changes (EFFCH) and technological changes (TECHCH) indices. EFFCH index remains unchanged during the study period with a value of 1.000. However, the TECHCH index shows different patterns during the study period. TECHCH index decreased in 2011-2012,2013-2014, 2016-2017 and 2020-2021 at the value of 0.595, 0.755, 0.878 and 0.572 respectively. The low TECHCH index was recorded during the transition period of the new policy or strategic plan implementation, especially during the 10th MP (2011-2012), DSTIN (2013-2014), 11th MP (2016-2017) and 12th MP (2020-2021). The low TECHCH index in 2011-2012 at the value of 0.595 can be translated to the impact of the Fukushima nuclear power plant accidents in March 2011. Whilst the low TECHCH index in 2019-2020 can be translated to the impact of the Covid-19 pandemic that affected all around the world in December 2019. The decreasing value can be translated to the ability of Nuklear Malaysia in adapting to the transition process of the new policy or programme implementation as well as the impact of global challenges such as the Fukushima nuclear power plant accidents that occurred in March 2011 and the Covid-19 pandemic. However, after the first year of implementation of the new policy or strategic plan, the TECHCH index increased to show an increasing value in 2012-2013 (7.0%), 2014-2015 (28.4%), and 2017-2018 (43.2%). During the implementation of the 12th MP 2016-2020, the average TECHCH index increased up to 5.16%, compared to the implementation of the 11th MP period which decreased to 12.34%.

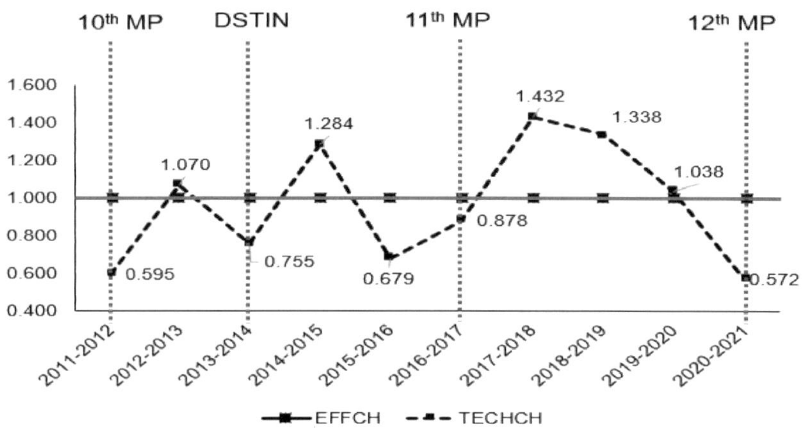

Figure 5. Efficiency changes and technological changes in nuclear technology R&D.

Fig. 6 shows the total factor productivity changes (TFPCH) index of the Malmquist Index of nuclear technology R&D in Nuklear Malaysia from 2011-2021. The nuclear technology R&D growth index in Nuklear Malaysia is mainly affected by technological changes due to policy changes compare to the resources provided since efficiency changes remain the same. The nuclear technology R&D growth index shows different patterns during the study period. The nuclear technology R&D growth index decreased in 2011-2012,2013-2014, 2016-2017 and 2020-2021 at the value of 0.595, 0.755, 0.878 and 0.572 respectively. The low nuclear technology R&D growth index was recorded during the transition period of the new policy or strategic plan implementation, especially during the 10th MP (2011-2012), DSTIN (2013-2014), 11th MP (2016-2017) and 12th MP (2020-2021). The nuclear technology R&D growth is also affected by major incidents such as the Fukushima nuclear power plant accidents that occurred in March 2011 and the Covid-19 pandemic that impacted the performance of R&D activities in Malaysia. The decreasing value can be translated to the ability of Nuklear Malaysia in adapting to the transition process of the new policy or programme implementation. After the first year of implementation of the new policy or strategic plan, the nuclear technology R&D growth index increased to show an increasing value of 2012-2013 (7.0%), 2014-2015 (28.4%), and 2017-2018 (43.2%). It can be concluded that the nuclear technology R&D growth index increased up to 5.16% during the implementation of the 12th MP 2016-2020 compared to the implementation of the 11th MP 2011-2015 which decreased to 12.34%.

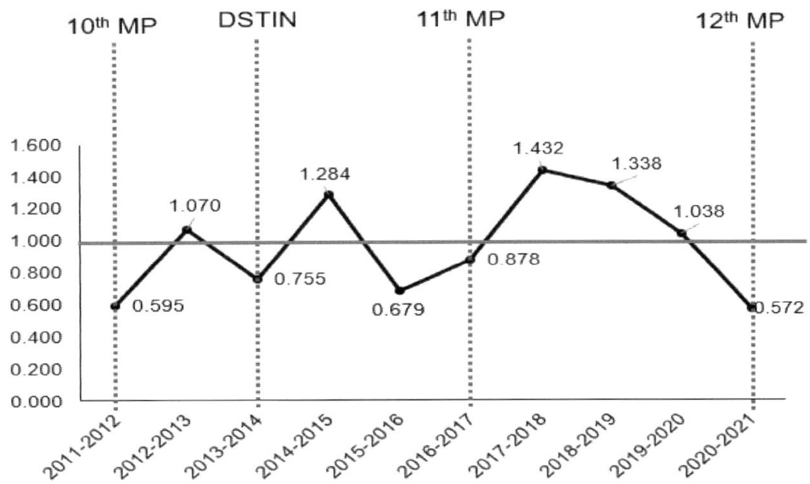

Figure 6. Total factor productivity changes (TFPCH) index of nuclear technology R&D.

4.3. Discussions

Any country's economic development in the future will be driven by technological innovation and application. There lies the significance of R&D because it is one type of spending that creates and propels these new technologies. Besides, R&D is crucial for improving the capacity of research institutes since it encourages the development of novel production techniques, lowers prices, and raises product quality. The importance of R&D among the factors directly contributing to the successful economic development of the emerging nations of the twenty-first century is becoming more and more clear daily. This is essential due to ongoing technological advancement, rival competitors, and constantly shifting consumer preferences, which would be impossible without an R&D programme. Several recommendations are put forth in light of the study's findings to ensure that research organizations such as Nuklear Malaysia continue to be pertinent to the country's socioeconomic development in nuclear technology R&D activities.

Innovation is known as a fundamental component of R&D activity that contributes to economic progress and a nation's social well-being. Therefore, at any level of growth, Nuklear Malaysia should place the highest focus on policies or strategic plans that encourage innovation. By developing a fundamental policy for nuclear technology research and development (R&D), the Ministry of Science, Technology, and Innovation (MOSTI) can encourage innovation and make sure Nuklear Malaysia can ensure the right returns on innovative investments. This ongoing R&D investment should be made possible by the continued development of emerging technologies like the internet of things (IoT), artificial intelligence (AI), machine learning (ML), blockchain, cybersecurity, and cloud computing. The benefits of investing in new technologies for both stakeholders and potential business partners are continuing to grow quickly along with the rate of development, promoting investment and research in a variety of sectors while the new technologies themselves will present new opportunities for innovation.

In addition, nuclear and isotopic techniques can be used as solutions to address new global challenges like climate change and the growing nutritional and medical needs of a growing global population as well as to support the expansion of industrialization for development. The IAEA's platform for scientific cooperation, research, development, and training should thus be seized as a chance to further the development of nuclear science and technology. The platform can be utilised to support sustainable development and the well-being of Malaysians, especially in the fields of food and agriculture, environmental protection, water management, industrial growth, and public health.

To facilitate technology transfer, which is another important factor for growth, it is necessary to deal with problems like information provision and capacity building. This is crucial for many developing nations, including Malaysia, which lacks the skills to identify where and how technology may be implemented. Even if the business environment is in decent form, a transfer of technology won't be optimal if there aren't enough engineers and scientists to do so. IAEA might therefore help Nuklear Malaysia establish relationships with organisations abroad to facilitate information flow and raise national awareness of current technology.

Technology foresight and road mapping can be employed as important tools and techniques to support the advancement of nuclear science and technology in national development and the transition to a sustainable future. Future improvements in nuclear science and technology will continue to benefit humankind's well-being together with continuous technological innovations and advancements. The complementary production elements required for innovation include competent employees, adequate equipment, funding, and managerial qualities. These components are crucial to the complementary production factors. The complementary production elements required for innovation include competent employees, adequate equipment, funding, and managerial qualities. These components are crucial to the complementary production factors. The complementary production factors required for innovation include competent employees, adequate equipment, funding, and managerial qualities. These factors are crucial to the complementary production factors. These factors are also crucial for finding and implementing new technologies as well as for conducting research and development in nuclear technology. Additionally, developing countries like Malaysia might not have the resources or human capital to carry out advanced R&D-focused activities, thus particular policies need to concentrate on these areas first. It has been demonstrated that managerial upgrading programmes, in which an outside expert evaluates an organization's performance and offers a development plan, may have a significant impact on productivity and innovation in the future.

5. Conclusions

This paper has examined the performance and the growth of nuclear technology R&D in Nuklear Malaysia from 2011-2021 using the DEA and MI analysis. The result shows that the performance of nuclear technology R&D in Nuklear Malaysia is 96.70 per cent from 2011-2021. The study shows that nuclear R&D growth decreased by 8.3 per cent for the entire study period. The technological changes decreased at 8.3 per cent, while the efficiency changes remain. The study also found that the nuclear technology R&D growth index in Nuklear Malaysia is mainly affected by technological changes due to policy changes compare to the resources provided since efficiency changes remain. The results also found that the nuclear technology R&D growth is affected by major incidents such as the Fukushima nuclear power plant accidents that occurred in March 2011 and the Covid-19 pandemic that impacted the performance of R&D activities in Malaysia. The findings of the study help to align the R&D institution's direction to ensure that nuclear technology R&D contribute to the socio-economic development of the country. There are a few other restrictions on this paper. It was carried out based on the case study of the Malaysian Nuclear Agency and does not represent the whole nuclear technology R&D activities in the country. A future study can be proposed to carry out the study which includes more research institutions and universities to evaluate the comprehensive nuclear technology R&D activities in the country.

6. References

[1] Arik M, Ndrianasy E R 2018 Federally funded R&D fuels regional economies: a panel data analysis *Glob. J. Account. Financ.* **2** 15–36. http://search.ebscohost.com/login.aspx?direct=true&db=buh&AN=133797426&site=ehost-live.

[2] Arik M, Arik H, Graves B 2019 Academic R &D across the states: Efficiency and its determinants, *Glob. J. Account. Financ.* **3** 31–55.

[3] Asmara I J, Achelia E, Simamora N G, Sartono B 2019 Measuring R&D Performance using data envelopment analysis (DEA): case Indonesia, *Int. J. Soc. Sci. Humanit.* **9** 91–96. https://doi.org/10.18178/ijssh.2019.v9.997.

[4] Belgin O 2019 Analysing R&D efficiency of Turkish regions using data envelopment analysis *Technol. Anal. Strateg. Manag.* **31** 1341–1352 https://doi.org/10.1080/09537325.2019.1613521.
[5] Bae Y, Chang H 2012 Efficiency and effectiveness between open and closed innovation: Empirical evidence in South Korean manufacturers *Technol. Anal. Strateg. Manag.* **24** 967–980 https://doi.org/10.1080/09537325.2012.724164.
[6] Hung C L, Shiu P J 2014 Evaluating project performance by removing external effects: Implications to the efficiency of research and development resource allocation, *Res. Eval.* **23** 366–380. https://doi.org/10.1093/reseval/rvu022.
[7] Chun D, Chung Y, Bang S 2015 Impact of firm size and industry type on R&D efficiency throughout innovation and commercialisation stages: evidence from Korean manufacturing firms *Technol. Anal. Strateg. Manag.* **27** 895–909 https://doi.org/10.1080/09537325.2015.1024645.
[8] Jang H, Lee S, Suh E 2016 A comparative analysis of the change in R&D efficiency: a case of R&D leaders in the technology industry *Technol. Anal. Strateg. Manag.* **28** 886–900 https://doi.org/10.1080/09537325.2016.1180354.
[9] Khoshnevis P, Teirlinck P 2018 Performance evaluation of R&D active firms, *Socioecon. Plann. Sci.* **61** 16–28. https://doi.org/10.1016/j.seps.2017.01.005.
[10] Qin X, Du D 2018 Measuring universities' R&D performance in China's provinces: a multistage efficiency and effectiveness perspective *Technol. Anal. Strateg. Manag.* **30** 1392–1408 https://doi.org/10.1080/09537325.2018.1473849.
[11] Xiong X, Yang G L 2018 Guan Z C Assessing R&D efficiency using a two-stage dynamic DEA model: A case study of research institutes in the Chinese Academy of Sciences, *J. Informetr.* **12** 784–805 https://doi.org/10.1016/j.joi.2018.07.003.
[12] Mohd Chachuli F S, Mat S, Ahmad N 2021 Performance evaluation of renewable energy R&D activities in Malaysia *Renew. Ener.* **163** 544–560 https://doi.org/10.1016/j.renene.2020.08.160.
[13] Li H, Chen C, Cook W D, Zhang J, Zhu J 2018 Two-stage network DEA: Who is the leader?, Omega (United Kingdom) **74** 15–19 https://doi.org/10.1016/j.omega.2016.12.009.
[14] Khoshnevis P, Teirlinck P 2018 Performance evaluation of R&D active firms, *Socioecon. Plann. Sci.* **61** 16–28 https://doi.org/10.1016/j.seps.2017.01.005.
[15] Emrouznejad A, Yang G L 2018 A survey and analysis of the first 40 years of scholarly literature in DEA: 1978–2016 *Socioecon. Plann. Sci.* **61** 4–8 https://doi.org/10.1016/j.seps.2017.01.008.
[16] Emrouznejad A, Banker R, Doraisamy S M, Arabi B 2014 Recent developments in data envelopment analysis and its applications, *Proc. 12th Int. Conf. DEA* (Kuala Lumpur, Malaysia) 1–437 https://doi.org/10.13140/RG.2.1.1649.3608.
[17] Liu J S, Lu L Y Y, Lu W M, Lin B J Y 2013 A survey of DEA applications, Omega **41** 893–902 https://doi.org/10.1016/j.omega.2012.11.004.
[18] Liu J S, Lu L Y Y, Lu W M 2016 Research fronts in data envelopment analysis, Omega **58** 33–45 https://doi.org/10.1016/j.omega.2015.04.004.
[19] Mohd Chachuli F S, Ahmad Ludin N, Md Jedi M A, Hamid N H 2021 Transition of renewable energy policies in Malaysia : Benchmarking with data envelopment analysis, *Renew. Sustain. Ener. Rev.* 150 111456 https://doi.org/10.1016/j.rser.2021.111456.
[20] Mohd Chachuli F S, Ludin N A, Mat S, Sopian K 2020 Renewable energy performance evaluation studies using the data envelopment analysis (DEA): A systematic review *J. Renew. Sustain. Ener.* **12** 1–15. https://doi.org/10.1063/5.0024750.
[21] Charnes A, Cooper W W, Rhodes E 1978 Measuring the efficiency of decision making units *Eur. J. Oper. Res.* **2** 429–444. https://doi.org/10.1016/0377-2217(78)90138-8.
[22] Cooper W W, Seiford L M, Tone K 1999 Data envelopment analysis: A comprehensive text with models, applications, references and DEA-solver software (Kluwer Academic Publishers)
[23] Zhou H, Yang Y, Chen Y, Zhu J 2018 Data envelopment analysis application in sustainability: The origins, development and future directions, *Eur. J. Oper. Res.* **264** 1–16 https://doi.org/10.1016/j.ejor.2017.06.023.
[24] Avkiran N K 2006 Productivity analysis in the service sector with data envelopment analysis, 3r

ed. (Australia: The University of Queensland).
[25] Charnes A, Cooper W W 1962 Programming with linear fractional functionals *Nav. Res. Logist.* **9** https://doi.org/10.1002/nav.10100.
[26] Banker R D, Charnes A, Cooper W W 1984 Some models for estimating technical and scale inefficiencies in data envelopment analysis, *Manage. Sci.* **30** 1078–1092 https://doi.org/10.1287/mnsc.30.9.1078.
[27] Woo C, Chung Y, Chun D, Seo H, Hong S 2015 The static and dynamic environmental efficiency of renewable energy: A Malmquist index analysis of OECD countries *Renew. Sustain. Energy Rev. 47* 367–376. https://doi.org/10.1016/j.rser.2015.03.070.
[28] Fare R, Grosskopf S, Norris M, Zhang Z 1994 Productivity growth, technical progress, and efficiency change in industrialized countries *Am. Econ. Rev. 84* 66–83 https://doi.org/10.2307/2117971.
[29] Caves D W, Christensen L R, Diewert W E 1982 The Economic theory of index numbers and the measurement of input, output, and productivity *Econometrica* **50** 1393 https://doi.org/10.2307/1913388.
[30] Menegaki A N 2013 Growth and renewable energy in Europe: Benchmarking with data envelopment analysis *Renew. Ener.* **60** 363–369 https://doi.org/10.1016/j.renene.2013.05.042.
[31] Nuklear Malaysia 2013 *Malaysian Nuclear Agency: Strategic Plan 2012-2020*
[32] MOSTI 2013 *Dasar sains, teknologi dan inovasi negara (DSTIN)* 2013-2020
[33] EPU 2010 Rancangan Malaysia ke sepuluh 2011-2015 (Putrajaya, Malaysia) https://www.pmo.gov.my/dokumenattached/RMK/RMK10_Eds.pdf.
[34] MEA 2018 Kajian separuh penggal rancangan Malaysia kesebelas 2016-2020: keutamaan dan penekanan baharu (Putrajaya, Malaysia) http://www.epu.gov.my.

Acknowledgements
The authors appreciate the support of the Planning and International Relations Division, Malaysian Nuclear Agency in this work.

IOP Conf. Series: Materials Science and Engineering 1285 (2023) 012007 doi:10.1088/1757-899X/1285/1/012007

Effect of particle size analysis on the physicochemical characterization of Malaysian monazite

Nurliana Roslan[1,a], Cik Rohaida Che Hak[1,b] and Abdul Rahman Norazumin[1,2,c]

[1]Malaysian Nuclear Agency, Bangi, 43000 Kajang, Selangor, Malaysia
[2]Faculty of Chemical Engineering, Universiti Teknologi MARA (UiTM), 40450 Shah Alam, Selangor, Malaysia

[a]nurliana_r@nuclearmalaysia.gov.my, [b]rohaida@nuclearmalaysia.gov.my, [c]abdrahman@nuclearmalaysia.gov.my

Abstract. Rare earth elements (REEs) are the elements that have become extremely important to our world of technology owing to their unique physical and chemical properties, and widely used in advanced technology. Monazite is one of the mineral phosphate that contain a high amount of REEs including thorium and a small amount of uranium. In Malaysia, monazite is exist as by-product (*'amang'*) from tin mining originated from Ipoh, Perak. The objective of this study is to determine the effects of the different particles size of raw and ground (milled) monazite on the physicochemical properties. The effectiveness of extraction and separation process of monazite was affected by the size of the particle. Based on the particle size distribution analysis, all the monazite samples were found to be present in the size range of 11-190 μm. The morphological and elemental composition of the monazite mineral have been characterized by field emission scanning electron microscope (FESEM) integrated with energy dispersive X-ray spectrometer (EDX). The FESEM micrographs of the monazite showed an irregular spherical structures with different sizes. The chemical composition and the mineral phases present in monazite were studied by wavelength-dispersive X-ray fluorescence spectrometer (WD-XRF) and X-ray diffraction (XRD). The XRF analysis confirmed the presence of the major REEs such as Ce, La and Nd elements in the mineral. XRD analysis also identified (Ce, La, Nd and Th)PO_4 as the major phases along with quartz as minor phase. According to the FESEM-EDX and XRF analysis, smaller particle size of monazite gave higher content of REEs compare to raw monazite.

1. Introduction

Rare earth elements (REEs) exist in a group of 17 from the lanthanide series as well as scandium and yttrium elements. These REEs can be classified into two category including light rare earth (LRE) and heavy rare earth (HRE) elements [1]. Currently, REEs are the elements that have become extremely important to our world of technology owing to their unique magnetic, phosphorescent and catalytic properties. The value of REEs is increasingly rising due to their uses in numerous modern technologies, including the applications as catalysts in refining petroleum and filter-neutralizers of exhaust gases of cars, permanent magnets (used in wind turbines and electronics), fiber optics, lasers, oxygen sensors, superconductors and rechargeable batteries [2].

Globally, the world reserves of REEs by principal countries such as China, Brazil, Vietnam, Russia and India, stand at about 130 million tonnes [3]. Among these countries, China represents the largest exporter and consumer of REEs at 105,000 tons with a mine productions rate of 83%, while Australia is the second largest producer with 11% mine production rate [4]. The value of REEs is increasingly rising due to their numerous applications in a variety of newly-developed technologies. However, due to the stringent exportation policy, China has drastically reduced the export of REEs causing increase in the REEs's price. As a result, demand for REEs has been increasing, thus the processes must improve to continue to meet this demand in the future. Therefore, Malaysia is the second Asian country after China, which enters the global supply chain of REEs production as well as securing supply in the REEs industry. In facts, monazite and xenotime are the minerals phosphates that exist as by-product of tin mining processing in Malaysia. `*Amang*` or tin tailing is a heavy mineral rejects which remain after tin oxide (cassiterite) has been extracted from concentrated tin ores [5]. Moreover, monazite and other minerals (xenotime, zircon and ilmenite) are separated from *amang via* physical separation such as flotation, gravity separation, electrostatic separation or magnetic processes [6].

Monazite mineral is one of the major sources for REEs including thorium and uranium. In particular, monazite contains a high amount of REEs which contribute about 54 – 60%, 3.1 – 11.34% of thorium and a small amount of uranium (0.2 – 0.4%) [7]. The REEs produced from monazite is a mixed product containing various individual elements. In particular, monazite mineral is chemically and thermally stable, thus it is difficult to digest either using alkali or acid solutions. Therefore, milling and sieving are the pre-treatment methods for size refinement of monazite. It is believed that this process will increase the efficiency of digestion and leaching reactions because its reduces the particle size of monazite.

In this study, we investigate the optimization of pre-treatment of Malaysian monazite particles using two different milling scale (laboratory and pilot plant). The effect of particle size distribution on the physicochemical properties of monazite will be characterized using Wavelength Dispersive X-ray Fluorescence Spectroscopy (WD-XRF), X-ray Powder Diffraction (XRD) and Field Emission Scanning Electron Microscopy with Energy Dispersive X-Ray Spectroscopy (FESEM-EDX).

2. Materials and methods

2.1. Sampling

Monazite mineral used in this study were originally obtained from *'amang'*, a by product from tin mining at processing factory in Ipoh, Perak. Monazite was undergo milling and sieving process as the pre-treatment methods for the particle size refinement.

2.2. Sample preparation

Optimization of the pre-treatment methods was carried out using two different milling scale (laboratory and pilot plant scale). Figure 1 shows the milling technique using planetary mono mills and roll ball milling machine. For laboratory scale, milling process was carried out using dry milling technique in planetary mono mills (Pulverisette 6, Fritsch) at 350 rpm for 15 - 20 minutes. In contrast, milling process in pilot plant scale was carried out using roll ball milling machine (MVLab) at constant rpm for 3 hours (30 minutes of interval time). After milling process, the ground monazite was sieved using vibratory sieve shaker (Fritsch) with different sized apertures. The abbreviations for monazite samples; before milling and after milling are tabulated in Table 1.

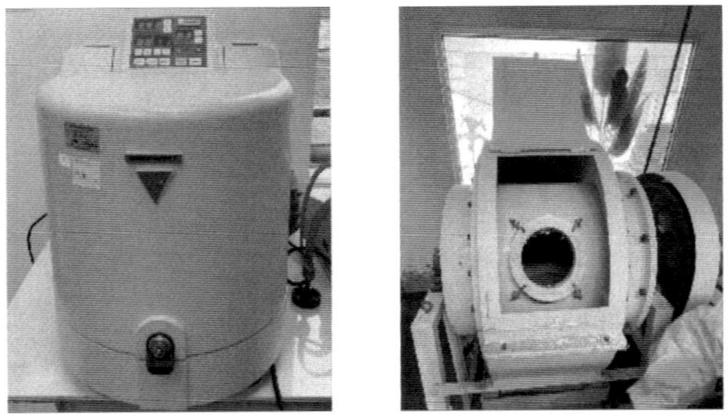

Figure 1. Milling technique (a) planetary mono mills (b) roll ball milling machine.

Table 1. Abbreviations of monazite samples.

Samples	Abbreviation
Raw monazite	RM
After milling monazite (laboratory scale)	MM-Lab
After sieving monazite (laboratory scale)	MS-Lab
After milling monazite (pilot plant scale)	MM-PP
After sieving monazite (pilot plant scale)	MS-PP

2.3. Characterization techniques

The representative portions of the monazite samples was prepared for preliminary analysis to determine the particle size distribution using particle size analyzer (Microtrac/X100, PSA). Quantitative analysis of the major and trace elements in the monazite samples was performed by wavelength-dispersive X-ray fluorescence spectroscopy (WD-XRF) using Rigaku (Model Supermini 200). A monazite samples of mass about 4-5 g was compressed into a pellet of 2.5 nm diameter, transferred into the sample chamber and then were bombarded with X-rays. The quantification analysis by WD-XRF was complemented with ZSX software. The identification of mineral present in the monazite samples was characterized using Panalytical, X'Pert Pro X-ray diffractometer (XRD). Morphological analysis of monazite samples were performed using field emission scanning electron microscope (FESEM, Carl Zeiss/ GeminiSEM 500) equipped with energy-dispersive X-ray spectroscopy (EDX, Oxford Instrument/ X-Max 80).

3. Results and Discussion

3.1. Physical observation of monazite

The physical observation of monazite samples for before milling and after milling and sieving is presented in Figure 2. It can be seen that the colour of RM looks like yellowish brown to reddish brown and exhibited a coarse-grained surface texture. After milling and sieving (MM-Lab), the colour of monazite looks like slightly greyish- yellow and the texture become smoother and looks like fine powder particles.

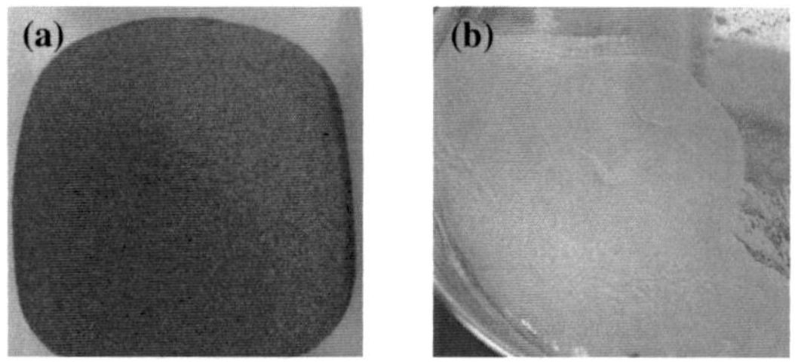

Figure 2. Monazite samples (a) before milling (RM) (b) after milling and sieving (MM-Lab).

3.2 Particle size analysis of monazite

Particle size analysis was conducted to study the size distribution of the monazite particles. Figure 3 shows the particle size distribution of the monazite samples. The understanding of the particle size distribution has a considerable impact on the efficiency of digestion and separation process.

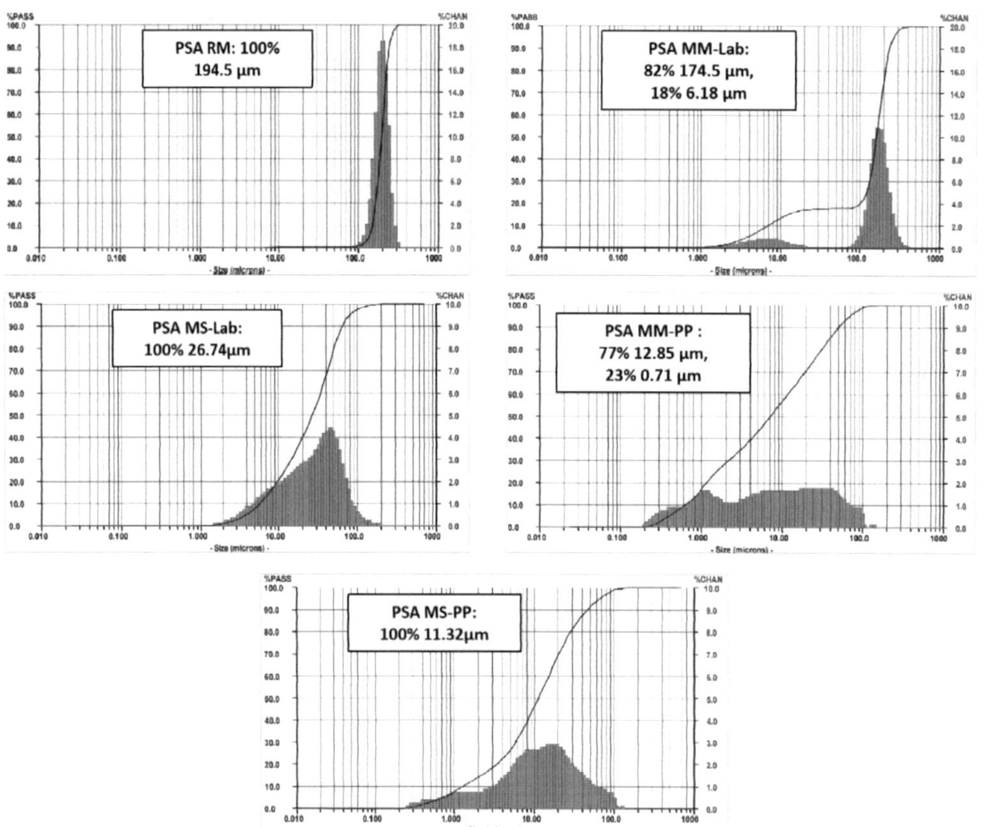

Figure 3. Particle size distribution of monazite samples

Figure 3 indicates that RM has single size distribution 194.5 µm in the range of 100 - 500 µm. As expected, the particle size of monazite decreases after milling and sieving process. The MM-Lab showed two peaks of distribution in the range of 1 - 15 and 100 - 500 µm with size of particles of 174.5 µm

(82%) and 6.18 µm (18%) respectively. It can be suggested that 15-20 minutes of milling monazite at 350 rpm (laboratory scale) may reduce some of the particle size but were not enough to produce homogeneous particles. After sieving process, the size of particles for MS-Lab sample became smaller with 26.74 µm (100%). In contrast, MM-PP also indicates two different groups of distribution with the particle size of 12.85 µm (77%) and 0.71 µm (23%), respectively. The size of monazite particle of MM-PP also decreases after sieving (MS-PP), which gave 11.32 µm (100%). These results suggested that sieving process does not give any significant changes of particle size when MM-PP sample was ground for 3 hours. Thus, milling and sieving process may resulted monazite with smaller size and lead to more homogeneous particles.

3.3. Effect of particle size analysis to the chemical composition of monazite

The chemical composition of the elements presents in the monazite samples conducted by WD-XRF analysis are presented in Table 2. The analysis indicated that an average of chemical composition accounted for rare earth elements in monazite are 50 wt.% out of the total composition. It can be seen that, the percentage of phosphorus correspond to the phosphate compounds which indicated that the monazite samples was significantly composed of rare earth phosphate minerals [5].

Table 2. Chemical composition of monazite samples by XRF.

Samples / Composition (%)	**RM**	**MM-Lab**	**MS-Lab**	**MM-PP**
Al	2.09	1.64	1.83	1.26
Si	2.53	3.34	2.86	2.74
P	16.23	17.43	17.87	19.10
S	0.27	0.18	0.21	0.25
Ca	0.95	1.36	1.32	1.36
Ti	1.48	0.72	0.82	0.44
Mn	0.17	0.00	0.00	0.00
Fe	3.46	1.65	1.78	0.84
Y	6.63	5.94	6.01	3.79
Zr	2.18	2.08	1.98	2.06
Sn	0.95	0.46	0.71	0.73
La	11.03	10.28	10.33	12.40
Ce	8.14	7.81	7.98	8.36
Pr	3.75	4.85	4.47	5.05
Nd	17.20	20.90	21.07	20.77
Sm	2.11	2.67	2.52	2.62
Gd	2.31	2.43	2.26	2.11
Dy	1.45	1.21	1.28	0.95
Er	0.88	0.52	0.59	0.35
Yb	0.79	0.46	0.41	0.34
Th	14.07	13.53	13.07	13.73
Others	0.27	0.17	0.19	0.17

The XRF results show that RM contained significant amount up to 50 wt. % of REEs, about 14.1 wt.% thorium (Th), 2 wt.% silicon (Si), 6 wt. % yttrium (Y) and 7 wt.% of other metals. These results also indicate that cerium (Ce), neodymium (Nd) and lanthanum (La) are the primary REEs in monazite. Other REEs including gadolinium (Gd), samarium (Sm), dysprosium (Dy), ytterbium (Yb) and erbium (Er) also presented in all monazite samples with low composition. On the other hand, as expected, the composition of light REEs such as Ce, Nd, La, Pr and Sm were more abundant in monazite compared to heavy REEs (Dy, Gd, Yb and Er). Apart from the REEs, a considerable amount of elements such as zirconium (Zr), iron (Fe) and titanium (Ti) was observed and could be the sourced to the accessory minerals occurring with monazite such as zircon, hematite and ilmenite or rutile, respectively. Table 2 shows that MM-PP sample contained the highest REEs composition compared to the other samples. This is because monazite is hard and dense mineral, therefore crushing the particles into smaller size through milling and sieving process could help monazite to expose more elements to the surface [8].

3.4. Effect of particle size analysis to the mineral identification of monazite

Generally, XRD is an important technique for identifying the mineral phases present in monazite minerals. The diffraction spectra of MS-Lab and MM-PP was shown in Figure 4. Based on the spectra, there are four (4) most significant peaks located around 25°, 27°, 28° and 31° that correspond to the mineral monazite-Ce, La, Nd and Th, respectively. It can be observed that MM-PP with clear diffraction peaks possessed a higher intensity compare to the MS-Lab. Moreover, both samples shows the typical peaks that corresponds to the rare earth phosphate of monazites minerals [5]. In view of this observation that is in agreement with the results of XRF analysis, it can be concluded that Malaysian monazite constitutes mainly of Ce, La, Nd and Th-phosphate and the present of Sm, Y, Gd, Al, Si-based compounds with low composition of [9].

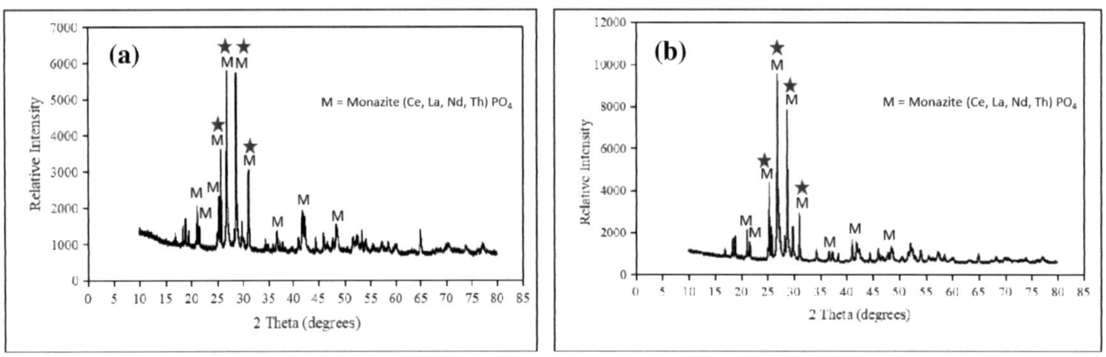

Figure 4. X-ray diffraction spectra of (a) MS-Lab and (b) MM-PP

3.5. Effect of particle size analysis to the morphology and elemental analysis of monazite

In order to study the microscopic morphology, monazite samples were further characterized by FESEM. Clear FESEM images of monazite samples were obtained at 88X magnification as shown in Figure 5. The micrographs of RM revealed that the particles are slightly spherical and elongated to sub-rounded in shape. Moreover, physical fractured features and irregular shaped particles are also observed on the surface of particles of MM-Lab, MS-Lab, MM-PP and MS-PP, respectively [10].

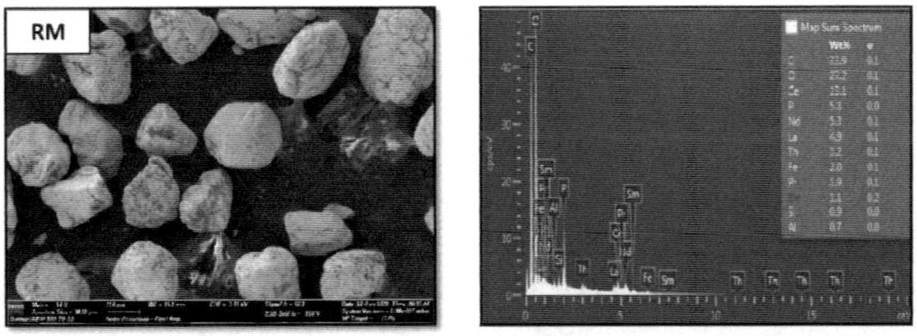

Element	Weight (Wt. %)
Ce	13.1
La	4.9
Nd	5.3
Th	3.2
P	5.8
Si	0.9
C	33.9
O	27.2

Element	Weight (Wt. %)
Ce	14.9
La	6.5
Nd	5.3
Th	3.5
P	10.4
Si	1.7
C	7.0
O	42.7

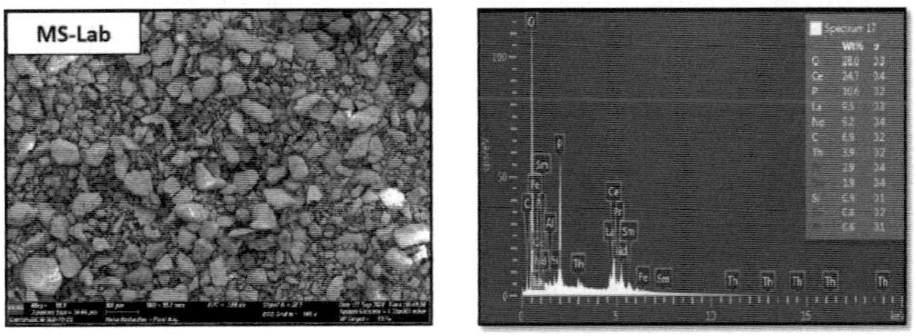

Element	Weight (Wt. %)
Ce	24.7
La	9.5
Nd	9.2
Th	3.9
P	10.6
Si	0.9
C	6.9
O	28.0

Element	Weight (Wt. %)
Ce	18.6
La	8.3
Nd	6.8
Th	4.6
P	9.9
Si	1.2
C	4.9
O	29.7

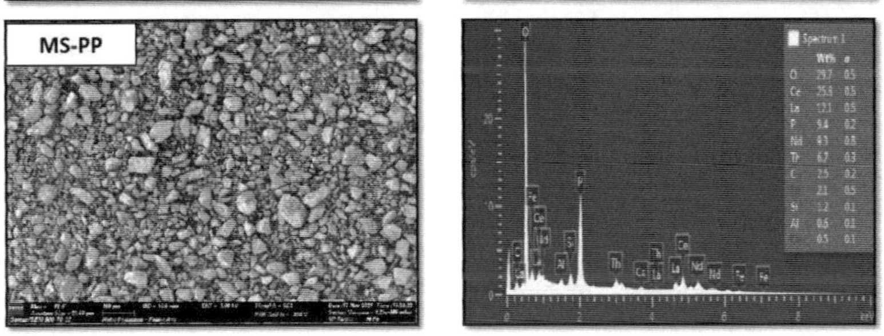

Element	Weight (Wt. %)
Ce	25.8
La	12.1
Nd	9.3
Th	6.7
P	12.1
Si	1.2
C	2.6
O	29.7

Figure 5. FESEM micrographs and EDX analysis of monazite samples.

These morphological features clearly reveal that the mechanical process of milling and sieving gave an impact on the surface of monazite particles. It is noted that these results are also in agreement with the particle size results which showed that milling and sieving process may reduce the particle size of monazite samples. Thus, it can be concluded that milling and sieving process may resulted monazite with smaller size but lead to more inhomogeneous morphology.

On the other hand, Figure 5 embedded with the table also shows the elemental composition of monazite before and after mechanical process using FESEM-EDX. The EDX analysis revealed the presence of the major elements of Ce, La, Th and Nd as phosphate. As can be seen from the results, there is an increase in the content of phosphorus of the monazite samples which was due to the exposure of the elements to the surface after milling and sieving process. Also, RM samples exhibited higher content of carbon composition due to the carbon tape in the poly-section. Of significance, the percentage content of Ce, La and Nd was increased in monazite samples compared to the RM, which is consistent with the chemical composition data obtained earlier from WD-XRF. In this study, it is conclusively proved that MM-PP displayed the highest weight concentration of REEs compared to other samples.

4. Conclusion

The pre-treatment method for the REEs recovery of the monazite samples was successfully optimized by milling and sieving process. Particle size analysis was carried out as a preliminary characterization to study the size distribution present in the monazite samples. The effect of particle size distribution of monazite on the physicochemical properties was studied using WD-XRF, XRD and FESEM-EDX analysis. Milling and sieving process resulted to a smaller size of monazite with inhomogeneous particles and these results are consistent with the surface morphological determined by FESEM. The combination of XRD analysis with WD-XRF and FESEM-EDX confirmed the presence of the major elements of REEs phosphate (Ce, La and Nd) and their composition (wt. %) in monazite mineral. In this study, pre-treatment of monazite using large milling scale (pilot plant) was chosen as an ideal sample for further chemical process because resulted in a smaller particles size and gave highest REEs composition compared to others.

5. References

[1] Voncken J H L 2015 Chapter 3: Physical and Chemical Properties of the Rare Earths *Rare Earth Elements: An Introduction (*Springer Briefs in Earth Sciences) 15-52.

[2] Charalampides G, Vatalis K I, Apostoplos B and Ploutarch-Nikolas B 2015 Rare earth elements: industrial applications and economic dependency of Europe *Procedia Econom. and Finance* **24** 126-135.

[3] Balaram V 2019 Rare Earth Elements: A Review of Applications, Occurrence, Exploration, Analysis, Recycling, and Environmental Impact *Geosci. Frontiers* **12** 1285-1303.

[4] Omodara L, Pitkäaho S, Turpeinen E M, Saavalainen P, Oravisjärvi K. and Keiski R L 2019 Recycling and Substitution of light rare elements, Cerium, Lanthanum, Neodymium, and Praseodymium from end-of-life applications- a review *J. of Cleaner Prod.* **236**(117573) 1-13.

[5] Udayakumar S, Mohd Noor A F, Sheikh Abdul Hamid S A R , Rama Putra T A and Anderson C G 2020 Chemical and mineralogical characterization of Malaysian monazite concentrate *Mining, Metallurgy & Explor.* **37** 415-431

[6] Kumari A., Panda R, Jha M K, Kumar J R and Lee J Y 2015 Process development to recover rare earth metals from monazite mineral: a review *Minerals Eng.***79** 102-115.

[7] Mohd Salehuddin A H J, Fazli Ismail A, Che Zainul Bahri, C N A and Aziman E S 2019 Economic analysis of Thorium extraction from monazite", *Nucl. Eng. and Technol.* **51** 631-640.

[8] Che Hak C H , Mahat S, Mustapha I and Sipaun S M Quantification of Thorium and rare earth elements in Perak's ex-mining monazite 2020 *Jurnal Sains Nuklear Malaysia* **32**(1) 10-19.
[9] Udayakumar S, Rezan S.A, Mohd Noor A F , Rama Putra T A , Mohamed Takip K and Hazan R, 2018 Characterization of Malaysian monazite concentrate for the recovery of Thorium Dioxide *J. of Phys. Conf. Ser*, **1082** 1-6.
[10] Anitha, J K, Joseph S, Rejith R G and Sundararajan M 2020 Monazite chemistry and its distribution along the coast of Neendakara-Kayamkulam Belt, Kerala, India *Springer Nature Appl. Sci.*. **2** (5) 1-18.

Acknowledgement

The authors gratefully acknowledge the Ministry of Science, Technology and Innovation (MOSTI) of Malaysia for the funding this project to purchase the MVLab Grinder through DSTIN Thorium Flagship Fund - Development of Innovative Nuclear Reactor Technology (with the Spin-Off) Based on Thorium (2015 - 2019). This research paper also was a part of the PQRD project (NM-R&D-21-77) entitled "Synergistic Solvent Extraction of Cerium from Monazite Mineral".

Feasibility study of Phosphor material synthesis process based on Barium halides via hydrothermal technique

Izura Izzuddin[1,a)], Siti Aishah Ahmad Fuzi[1] and Sapizah Rahim[1]

[1] Malaysian Nuclear Agency, Bangi 43000 Kajang Selangor, Malaysia

[a]izura@nuclearmalaysia.gov.my

Abstract. Barium halides based was typical host for phosphor material applied as imaging plate for computed radiography application. In this study, the host phosphor materials based on barium halides has been synthesized via hydrothermal technique. The aim of this study is to determine the feasibility of hydrothermal technique to synthesis phosphor materials. Three samples have been prepared for this study. All samples had fix Ba and Br composition but was differentiated by the third composition i. e F, Cl and I. X-ray diffraction (XRD) technique has been used to determine the crystallinity of the prepared samples while optical property was observed using photoluminescence (PL) spectroscopy. The sharp, narrow and high intensity peaks from XRD diffractogram showed that all samples have good crystallinity. The PL spectrum indicated that the emission of all samples corresponds to blue range. This study proved that hydrothermal method was feasible for phosphor materials synthesis process based on barium halides and offered lower temperature with shorter period of heating process which is cost effective for synthesis process.

1. Introduction

Phosphor material referred to material that released stored energy as luminescent signal upon stimulation of visible light. This unique property has made phosphor a prime material for imaging plate fabrication for computed radiography (CR) technology. Barium halides based doped with activator such as BaBrI [3], BaBrCl:Eu [11], BaBrF:Eu [4] have been widely studied and recognized as one of the established phosphor materials that are available in the market for imaging application. Previously, the most common technique that has been applied to synthesis this alkaline halides based was conventional sintering process [1][2]. Practically this method was tedious because required high temperature with multiple steps of heating hence consumes long period of time. The other techniques applied later on were Czochralski method [5] and the most familiar was Bridgman technique [7][1]. Even though the techniques are well known could improve the quality of luminescence properties, the techniques still operated at high temperature.

Recently, hydrothermal is one of the techniques that has attracted researcher to synthesis materials at low temperature especially for nano materials. However, this technique is rarely used for barium halides synthesis technique. In this study, this technique will be employed to synthesis barium halides based as an optional to other available method. As a starting, only host material will be synthesized. The aim of this preliminary study is to explore the feasibility of this technique to synthesize barium halides based.

2. Experimental method

Alkaline earth halides were synthesized via hydrothermal method. The raw materials used were Ba(NO2), NH_4Br, NH_4F, NH_4Cl and NH_4I powders with 99.99% purity. All the starting materials were individually diluted in distilled water to form precursor solutions. The solutions were stoichiometry mixed and continued with stiring process for 12hrs. The mixed solutions were transferred to teflon - lined closed vessel autoclave and hydrothermally treated in furnace for 30 minutes at 160 °C. Finally, the solutions were dried at 160°C for 2hrs to form the powders. Three different samples has been prepared for this experiment namely BaBrF, BaBrCl and BaBrI. Every sample consist of similar content of alkaline earth element i.e barium (Ba) and first halides element i.e bromin (Br) but different composition of second halides elements i.e fluorine (F), clorine (Cl) and iodine (I). Pananalytical X'pert Pro X-ray diffractometer has been used to determine the crystal properties of the samples while optical properties was observed using Edinburgh Instrument FLS920 photoluminescence (PL) spectroscopy

3. Result and discussion

The XRD result of prepared samples were depicted in Figure 1. In general, all spectrum of the prepared samples appeared sharp, narrow with high intensity peaks signified that all samples have good crystallinity.

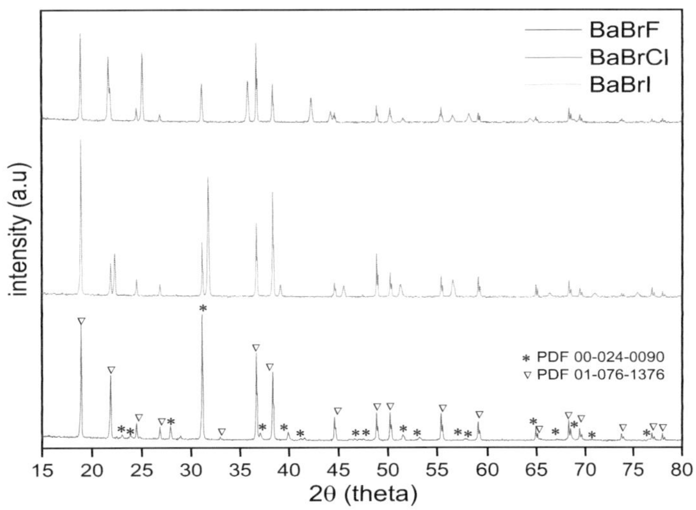

Figure 1. XRD spectrum.

The matched peaks procedure with database using eva software were done to identified the synthesis materials and their crystal structure properties. For BaBrF sample, all peaks were matched with PDF 00-024-0090 file of BaBrF and PDF 01-076-1376 file of $Ba(NO_2)_3$ as shown in Figure 2. The identical BaBrF spectrum with PDF 00-024-0090 file of BaBrF database demonstrated the feasibility of hydrothermal technique to synthesize phosphor materials. The synthesized BaBrF has tetragonal structure with space group P4/nmm(129). In addition, the matched peaks with $Ba(NO_2)_3$ database indicated no other impurities detected for this sample except $Ba(NO_3)_2$. In comparison, the ratio between BaBrF: $Ba(NO_3)_2$ was ~ 48:52.

IOP Conf. Series: Materials Science and Engineering 1285 (2023) 012008 doi:10.1088/1757-899X/1285/1/012008

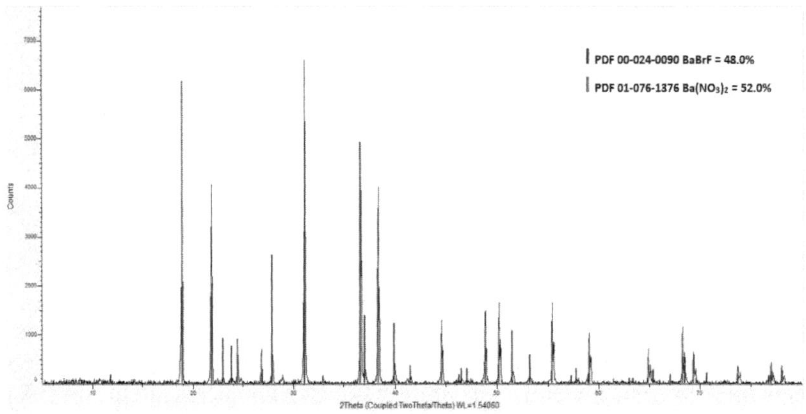

Figure 2. Peaks matched of XRD spectrum.

$Ba(NO_3)_2$ that has been detected was the precursor to prepare all samples. Huge amount of $Ba(NO_3)_2$ as impurity denoted that the applied hydrothermal technique need a lot of emendation for better yield. It is possible that the hydrothermal process was incomplete subsequently it ended up with the precursor as one of the final products. The cause of this circumstances seems most likely due to insufficient period for the chemical's reaction. Hence, one of the approaches that should be considered is to prolong the period of the hydrothermal process. Another parameter that might affected the final product was the pH of the precursor solution. In this study, the precursor solution has the pH value of 5. Past research reported that variation of pH influenced the crystal structure growth yet affect the optical properties of the prepared sample [9][12].

Due to limited database for phosphor materials, neither BaBrCl (Figure 1(b)) nor BaBrI (Figure 1(c)) samples were matched with available database. At first glance, both samples showed comparable pattern as BaBrF because almost all peaks that were not matched with $Ba(NO_3)_2$ appeared at the same positioned as BaBrF. With fixed composition of Ba and Br with similar synthesis technique, the chances for Cl or I to replace the same positioned of F in the BaBrF crystal structure system was high. This assumption was consolidated because Cl, I and F elements were clan of halides. Accordingly, they tend to have similar atomic arrangement in crystal structure but with different lattice parameter and reflected as comparable XRD spectrums. Furthermore, huge amount of $Ba(NO_2)_3$ as impurity also detected for both BaBrCl and BaBrI samples. The presence of $Ba(NO_2)_3$ was expected since the samples was followed the same routes of BaBrF sample preparation. To improve the quality of yield, suggested procedure as BaBrF will be repeated for these samples.

Another additional information can be measured from XRD result was crystallite size using Scherer equation. The highest peak at positioned ~31 has been chosen for fwhm value. The calculated crystallite size for BaBrF, BaBrCl and BaBrI were 55nm, 49nm and 50 nm respectively. The crystallite size generally depended on the fwhm of the chosen peak thus it has no direct relation with grain size or particle size of the samples.

Figure 3 shows the PL emission spectra of prepared samples. In general, all samples showed similar spectrum profile but differ in terms of spectrum intensity. This result was expected because it was in line with XRD result. Indirectly it indicates that crystal structure affects the luminescence properties of the sample. The emission peaks for all samples appeared in blue range wavelength using a 300 nm excitation wavelength. These profiles were typical emission profile of alkaline earth halides that has been reported by other studies simultaneously demonstrated the feasibility of hydrothermal technique to synthesis phosphor material. In this study, all samples showed their main peaks at 470nm (2.61 eV) [2]. The other peaks that belong to phosphor materials were detected at 450nm (2.76 eV) [8].

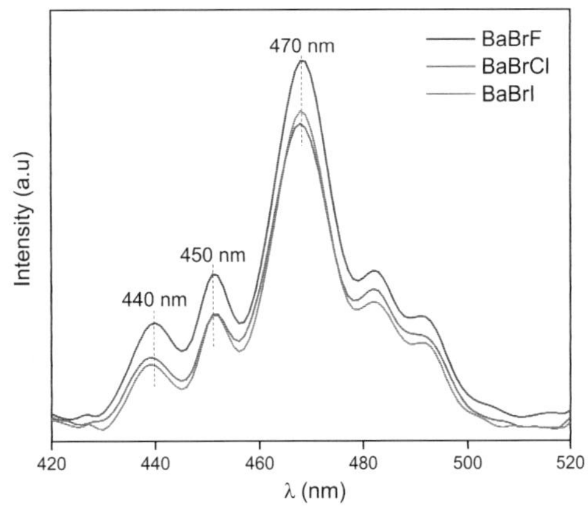

Figure 3. PL spectrum.

The observed peaks could be assigned to F-centers vacancies trapped electron. Two different types of F-centers were expected for every sample contributed by two anions from halides that existing in the samples. Therefore, it was confirmed that the F-centres for all samples were related to F(Br). Another F-centers for BaBrF, BaBrCl and BaBrI belong to F(F), F(Cl) and F(I) respectively. Additionally, the emission peaks also occured due to self-trapped exciton (STE) emission [8]. Meanwhile, the variation of profile spectrum intensity between all samples can be correlate with the ionization properties of the different halide elements. It is possible that sample with highest ionization properties will easily get excited and reflected as higher emission. Accordance to degree of ionization, F has better emission properties followed by Cl and I element and this sequence was proportionate with BaBrF, BaBrCl and BaBrI emission intensity.

Besides that, another significant emission peaks positioned at 440nm and was discovered. These peaks were declared as $Ba(NO_3)_2$ impurity [6]. These peaks were also found being in blue range spectrum and this implies more trap centre for the system. By modified the synthesis method as mentioned above these peaks were expected to be reduce later. In contrast, the main phosphor materials peaks need to be increased and can be achieved by addition of activator. One of the common activators is Eu element

4. Conclusion

From this preliminary study, the hydrothermal technique was feasible to synthesis phosphorous materials. In comparison with other typical techniques such as normal sintering, Czochralski and Bridgman techniques, hydrothermal technique offers lower temperature and shorter period of heating process which in turn was cost effective for synthesis process. However, to ensure these materials can be used as proper phosphor material for certain application, a lot of improvement need to be done. Further experiments will emphasize on modification of the preparation parameter to reduce the amount of impurity. Furthermore, this study will be proceeded with addition of Eu as an activator to enhance the luminescence properties of the prepared sample.

5. References

[1] Beck H P 1978 New mixed halide compounds MFX of divalent rare earths (M = Sm, Eu, Tm, and Yb; X = Cl, Br, and I) *J. of Sol. State Chem.* **23**(1–2) 213–217.

[2] Hesse S, Zimmermann J, Von Seggern H, Meng X, Fasel C, Riedel R 2009 Synthesis and functionality of the storage phosphor BaFBr: Eu^{2+} *J. of Appl. Phys.* **105**(6).

[3] Lenus, A J, Sornadurai D, Govinda Rajan K, Purniah B 2002 Luminescence beaviour of Eu^{2+}-doped BaClI and BaBrI *Mater. Lett.* **57**(3) 635–638.

[4] Ohuchi H, Yamadera A 2002 Possible application of an imaging plate to space radiation dosimetry *J. of Radiat. Res.* **43** 74–77.

[5] Rupasov A, Shalaev A, Shendrik R 2020 Growth of BaBrI crystals by the Czochralski method *Cryst. Growth and Des.* **20**(4) 2547–2552.

[6] Selvi R, Anand S, Durga R R, Sundararajan R S 2022 Growth , structural , spectroscopic , thermal and non-linear optical studies of L-Arginine Barium Nitrate crystals *J. of Emerg. Technol. and Innov. Res.* **9**(5) 787–795.

[7] Shalaev A A, Shendrik R, Myasnikova A S, Bogdanov A, Rusakov A, Vasilkovskyi A 2018 Luminescence of BaBrI and SrBrI single crystals doped with Eu^{2+} Optic. Mater. **79** 84–89.

[8] Shendrik R, Myasnikova A, Rupasov A, Shalaev A 2019 Role of electron and hole centers in energy transfer in BaBrI crystals *Radiat. Meas.* **122** 17–21.

[9] Shi X, Liu, Y, Zhang J, Zhang K, Li P, Zuo H, Li J 2015 Effects of pH and Sm^{3+} doping on the structure, morphology and luminescence properties of $BiPO_4$:Sm^{3+} phosphors prepared by hydrothermal method *Ceramics Int.* **41**(2) 3162–3168.

[10] Wolszczak W, Krämer K W, Dorenbos P 2020 Engineering near-infrared emitting scintillators with efficient $Eu^{2+} \rightarrow Sm^{2+}$ energy transfer *J. of Lumines.* **222**.

[11] Yan Z, Shalapska T, Bourret E D 2016 Czochralski growth of the mixed halides BaBrCl and BaBrCl:Eu. J. of Cryst. Growth **435** 42–45.

[12] Yu X, Wang Z, Wang Q, Mi X 2022 Warm white emission of $Ca_2P_2O_7$: Dy^{3+}, Eu^{3+} phosphor via hydrothermal method *J. of Alloys and Comp.* **897** 162745.

IOP Conf. Series: Materials Science and Engineering 1285 (2023) 012009 doi:10.1088/1757-899X/1285/1/012009

Effects of gamma radiation on sheep wool fiber composites reinforced with rice husk ash

Norsyahidah Mohd Hidzir[1,2,a)], Afifah Mardhiah Mohamed Radzi[1,2] and Nur Ain Mohd Radzali[1,2]

[1]Department of Applied Physics, Faculty of Science and Technology, Universiti Kebangsaan Malaysia (UKM), 43600 Bangi, Selangor, Malaysia
[2]Nuclear Technology Research Centre, Faculty of Science and Technology, Universiti Kebangsaan Malaysia (UKM), 43600 Bangi, Selangor, Malaysia

[a]syahidah@ukm.edu.my

Abstract. It is well known that animal fiber is one of the potential pollutants to the environment if not been well disposed. Besides animal fiber, rice husk ash has also been considered as a type of waste that are able to pollute the air and soil. Recently, there are plenty of research for the purpose of reusing this type of waste in the composite making. It has been established that irradiation on composite at dose of more than 3 kGy can potentially increase the properties of the composite as it becomes harder and able to hold huge loads. Therefore, this research is conducted to study the effect of irradiated composites of sheep wool fibers reinforced with rice husk ash. The composites were prepared with different percentage of sheep wool (3% and 5%) as well as the addition of rice husk ash as filler. The gamma irradiation of 7 kGy of dose is used on the composites and later analysed for the changes in the chemical and physical properties. New chemical bonds were observed for irradiated composites through Fourier-transform infrared spectroscopy (FTIR) analysis. Irradiated composite with 5% of wool has higher strength (0.16 MPa) compared to the irradiated composite with 3% of wool (0.09 MPa) from Izod fracture test. Meanwhile, composite without rice husk ash added were measured to have higher strength compared to the addition of rice husk ash as filler for both unirradiated and irradiated samples. The addition of rice husk ash increased the brittleness of the composite. In conclusion, the sheep wool fibers without the addition of rice hush ash are suitable to be used as reinforcing fibers in composites with gamma irradiation play insignificant part of the composite's strength.

1. Introduction

Natural fibers have many advantages over synthetic fibers in terms of low production costs [1], easy to decompose, low density, high strength, and low density [2]. Meanwhile, animal fiber or better known as keratin is a type of fibrous protein fiber that is found in sheep's wool, nails, and frond wool. Keratin differs from other protein fibers because keratin has a structure built from the amino acid cysteine that distinguishes it from other fiber proteins in terms of physical and thermal stability [2]. Therefore, this keratin waste needs to be disposed of or reused as an ingredient in the manufacture of products based on natural waste products. It is well known that animal fiber is one of the potential pollutants to the environment if not been well disposed.

Besides animal fiber, rice husk ash has also been considered as a type of waste that are able to pollute the air and soil. Rice husk is the residue of rice milling while rice husk ash is produced from rice husk burning. The burning of rice husks has been found to have produced rice husk ash containing up to 80-90% silica crystals [3]. In a study by Saravanan & Kumar (2013) and Fernandesa et al. (2018) have concluded that the tensile strength, compressive strength, and hardness increase in parallel with the increase in the mass fraction of rice husk ash in aluminum alloy and epoxy composite, respectively [4][5].

Composites are produced from a combination of polymer matrix and reinforcing fibers. The fibers used in the manufacture of composites are made up of synthetic or natural materials. Hybrid composites can also be produced through a combination of natural and synthetic materials [6]. The purpose of composite production is to produce new materials with high strength and to overcome the shortcomings of the materials used [7].

Irradiation on composites is one of the techniques that can promote chemical reactions. Composites will undergo cross-linking or chain degradation when irradiated depending on the dominant process during dose absorption. Previous study has found the irradiated composite undergo oxidation as well as the addition of functional groups [8].

Therefore, this research is conducted to observe the effect of irradiation and sheep wool reinforcement fiber composite on strength, chemical and physical properties through mechanical test, as well as chemical and physical changes from various instrument such as Fourier-transform infrared spectroscopy (FTIR), Scanning electron microscope (SEM), and Izod fracture test. In addition, the used of rice husk ash as filler is investigated. This study will also determine the appropriate percentage of animal fibers in the production of the best animal fiber composites.

2. Materials and Methods

Epoxy resin was obtained from Ezlazer Malaysia and sheep wool fiber was obtained from Spotlight Stores Sdn. Bhd. The hand lay-up technique was used in the production of composite (epoxy resin and sheep wool fiber). Composites are prepared using different percentages of sheep wool fiber (3% and 5%). In addition, rice husk ash is added as a filler into the composite. The samples were irradiated with gamma irradiation at 7 kGy using a GMS Nordion Gamma Cell model GC 200 E with a dose of 0.713 kGy/hour. Changes in the composites were chemically and physically characterized for comparison with non-irradiated control composites using Scanning Electron Microscope (SEM), Fourier Transform Infrared Spectroscopy (FTIR) and Izod fracture test.

SEM analysis was carried out using a Zeiss Merlin SEM machine using 10 kV at various magnification.

For FTIR analysis (Perkin-Elmer model Spectrum 400FT-IR/NIR with Imaging System), composite samples were measured in the range 650-4900 cm^{-1} at a resolution of 8 cm^{-1} in approximately 128 scans.

The Izod fracture test (Model Shimadzu) was performed on all irradiated and unirradiated samples. Each sample with dimensions of $10 \times 2 \times 1.2$ cm will be notched or bent before being subjected to the same load of 0.449 kg at a velocity of 3.5 ms^{-1}. The applied load is 2.75 N. The results of this test are calculated in MPa.

3. Results and Discussion

Figure 1 shows the color changes for irradiated composites (3% and 5% fiber) from crystal clear to yellowish due to the oxidation that occurred.

In Figure 2(b) and 2(d), SEM analysis shows that the irradiated composites (with 3 and 5% wool fiber) have a strong adhesion between the fibers and the matrix. This could indicate the increased of interfacial bonding in the composite from gamma radiation.

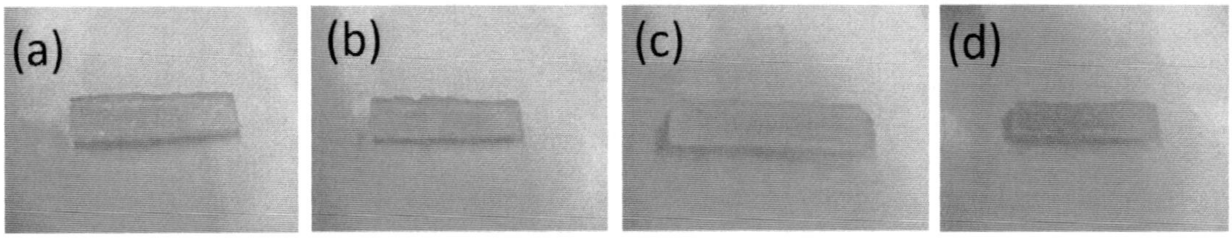

Figure 1. Observations on sheep wool fiber composites a) 3% fiber control sample b) 3% fiber sample irradiated at 7 kGy c) 5% fiber control sample, and d) 5% fiber sample irradiated at 7 kGy.

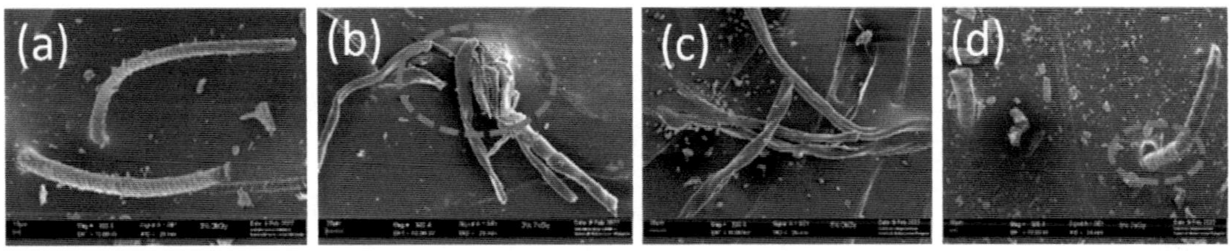

Figure 2. SEM analysis of sheep wool fiber composite a) control sample 3% fiber b) sample 3% fiber irradiated at 7 kGy c) control sample 5% fiber d) sample 5% fiber irradiated at 7 kGy.

In Figure 3, FTIR analysis of the composite containing 3% animal fiber that has been irradiated as well as the composite added with rice husk ash show significant band representing the presence of amine functional groups at 1050-1580 cm^{-1} as well as carboxyl groups. The presence of abundant amine functional groups indicate that amine functional groups are present in sheep wool as well as epoxy resin. The bending also shows the presence of hydroxyl groups for irradiated composites (Figure 3(b) and 3(d)) at 3300-3400 cm^{-1} resulting from the oxidation that occurs during irradiation. In Figure 4, the bend that represents the bond of the amine functional group were also observed. The presence of carboxyl groups can also be clearly seen in composites containing rice husk ash, similar to Figure 3. Through a previous study by Xu et al. 2007, irradiation on fiber and matrix grafts causes the presence of double oxygen where this situation is explained from the addition of bonds involving large functional groups that contain oxygen, namely hydroxyl and carboxyl groups in current study [8].

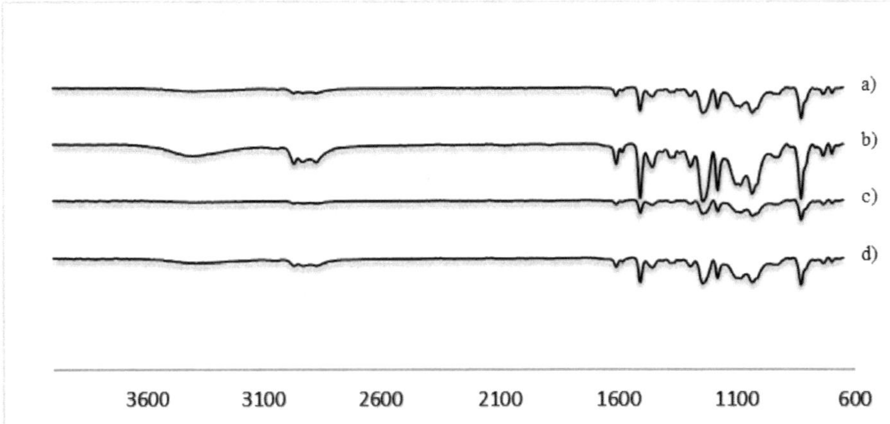

Figure 3. FTIR analysis of 3% sheep wool fiber composite a) control sample, b) sample irradiated at 7 kGy, c) control sample containing rice husk ash, and d) sample containing rice husk ash and irradiated at 7 kGy.

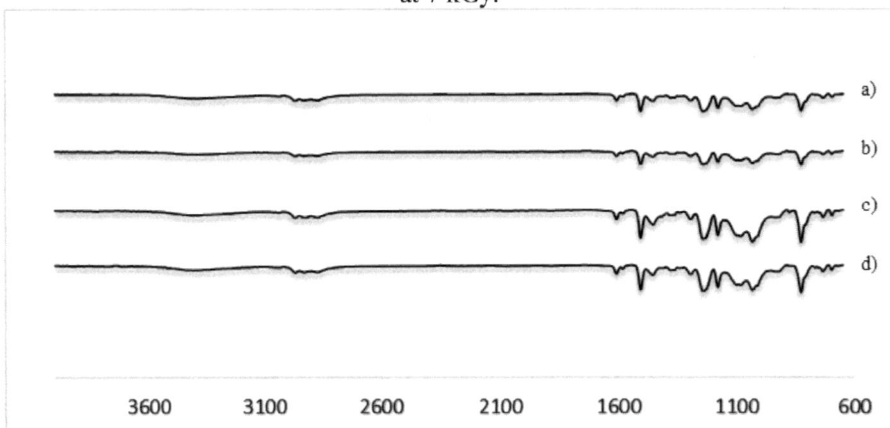

Figure 4. FTIR analysis of 5% sheep wool fiber composite a) control sample, b) sample irradiated at 7 kGy, c) control sample containing rice husk ash, and d) sample containing rice husk ash and irradiated at 7 kGy.

The Izod test (Table 1) shows that the strength of the irradiated composite is almost similar to non-irradiated composite. Non-irradiated and irradiated composite with 5% of wool (without the addition rice husk ash) has higher strength (0.161 MPa) compared to the composite with 3% of wool (without the addition rice husk ash) (~0.1 MPa) from Izod fracture test. This study is in line with the previous study reported by Verma et. al. (2018) with 3.01 MPa was observed in that study for the composite produced containing 5% animal fiber compared to 2.67 MPa for the composite containing 3% animal [9]. In addition, the results of the Izod fracture test in Table 1 shows that the addition of rice husk ash in the composite has increased the fragility of the composite. Composite without rice husk ash added were measured to have higher strength compared to the addition of rice husk ash as filler for both unirradiated and irradiated samples. This indicates that the addition of rice husk ash increased the brittleness of the composite.

Table 1. Izod fracture Test results for composite samples.

Sample	Average energy absorved ± standard deviation (Mpa)	
	a) 0 kGy	b) 7 kGy

3%	0.102 ± 0.011	0.099 ± 0.023
3% + rice husk ash	0.100 ± 0.031	0.083 ± 0.017
5%	0.161 ± 0.053	0.161 ± 0.048
5% + rice husk ash	0.058 ± 0.008	0.059 ± 0.010

4. Conclusion

In conclusion, the sheep wool fibers without the addition of rice hush ash are suitable to be used as reinforcing fibers in composites with gamma irradiation play insignificant part of the composite's strength. The composite with higher percentage of wool (5%) is better than 3% of wool as the strength showed good agreement with the increased of fiber percentage in the composites. Future work with higher percentage of fiber (greater than 5%) could possibly results in a better mechanical characteristic of the composites.

References

[1] Mohammed L, Ansari M N M, Pua G, Jawaid M and Islam M S 2015 *Int. J. Polym. Sci.* **2015** 1-15
[2] Kim N K, Lin R J T and Bhattacharryya D 2014 *Compos. Part B Eng.* **67** 472–480
[3] Nagrale S D, Hajare H and Modak P R 2012 *Int. J. Eng. Res. Appl.* **2** 1–5
[4] Saravanan S D and Kumar M S 2013 *Procedia Eng.* **64** 1505–13
[5] Fernandesa I J, Santos R V, Dos Santos E C A, Rocha T L A C, Junior N S D and Moraes C A M 2018 *Mater. Res.* **21** 1-10
[6] Amir S M M, Sultan M T H, Jawaid M, Safri S N A, Shah A U M, Yusof M R, Naveen J, Mohd S, Salleh K A M and Saba N 2019 *J. Mater. Res. Tech.* **8** 5362–73
[7] Park S J and Seo M K 2011 Types of Composites *Interface Science and Technology* (Electronic Materials vol 18) ed Park S J and Seo M K (The Netherlands: Elsevier) chapter 7 pp 501-629
[8] Xu Z, Huang Y, Zhang C, Liu L, Zhang Y and Wang L 2007 *Comp. Sci. Tech.* **67** 3261–70
[9] Verma A, Negi P & Singh V K 2018 *Polym. Comp.* **40** 1–10

Acknowledgments

This study was conducted using research grant FRGS/1/2019/STG02/UKM/02/6 from the Malaysia Ministry of Higher Education. Acknowledgements are given to the Nuclear Science Laboratory for gamma irradiation, the Center for Research and Instrumentation Management (CRIM), Universiti Kebangsaan Malaysia (UKM) for SEM and FTIR analysis and the Materials Science Laboratory for Izod Fracture Testing.

Gamma radiation shielding simulation for thermoplastic elastomer and silicon rubber/alloy dopant with functional nanoparticle

Chai Chan Ful[1], Muhammad Arif Sazali[1,a)], Khairulnadzmi Jamaluddin[1], Nur Syazwani Mohd Ali[1], Muhammad Syahir Sarkawi[1]

[1]School of Chemical and Energy Engineering, Faculty of Engineering, Universiti Teknologi Malaysia, 81310 Skudai, Johor, Malaysia

[a]muhammadarifsazali@utm.my

Abstract. With the global energy crisis worsening, the research and application of new energy have become a strategic objective for the major countries' long-term development in the twenty-first century. Nuclear energy technology development and application have become central to new energy development. Numerous types of high-energy radiation (such as X-rays, gamma rays, and neutron rays) produced by nuclear materials not only impose stricter requirements on the radiation-resistant materials used in the reactor but also risk causing significant environmental pollution (nuclear pollution) and serious harm to human life and health if a leak does not form. As a result, the prudent control and protection of nuclear radiation have become critical issues in the development of nuclear energy. With the rapid development of nuclear power today, it is necessary to seek new nuclear radiation protection materials, particularly those with high energy, penetrating ability, and other neutron-protective material characteristics. This is a new type of nuclear power design and construction of critical materials. In this study, WinXCom software is used to simulate the shielding properties of the used functional shielding element, compounds of metal oxide, and the mixture of RTV and TPE with the functional nanoparticle and observe the K-edge absorption of the material used. Based on the simulation, the shielding performance of the selected element of Sm was observed to have the smallest weak gap absorption zone of 40-47 MeV in common medical radiation of 30-150 KeV. Different sets of combinations of shielding material were simulated and found that the combination of WO_3 with Sm_2O_3 gave the most pronounced radiation shielding performance with two strong absorption peaks at 46.8 KeV and 69.5 KeV. RTV composite (Sm_2O_3+RTV+ WO_3) with WO_3 as matrix gave the best shielding performance and was used as a reference for the manufacture of the new type of shielding materials.

1. Introduction

In recent years, new alloys, structures, shapes, and other materials have emerged. A. B. Azeez [1] used natural rubber as the base and iron powder as the functional filler to prepare lead-free γ-ray radiation protection materials with low radiation shielding performance. The Gd_2O_3/rubber flexible leadfree shielding composite materials studied by Mao *et al.* [2] the BO/PDMS flexible lead-free shielding composite materials developed by S. Nambiar *et al.* [3], and the W/SEBS flexible leadfree shielding composite materials developed by Luo *et al.* [4]. The shielding effect of neutrons is the slowing down of fast neutrons and the absorption of slow neutrons.

In recent years, due to flexible shielding materials with flexible, small specific gravity, can be arbitrarily bent, repeatedly folded, cut, and easy to use other characteristics, especially in some equipment that cannot deal with holes and gaps. A nanoparticle is a small particle that ranges between 1 to 100 nanometers in size. Undetectable by the human eye, nanoparticles can exhibit significantly different physical and chemical properties to their larger material counterparts. Fu Ming *et al.* [5] researched a flexible butadiene rubber-based shielding material through a blending process, using reasonable multi-component additives to improve the physicochemical properties of the rubber, which is not only arbitrarily deformable, flexible, and has good shielding properties with structural adaptability and simple processing. Furthermore, they still add traditional lead which causing difficulty to produce high quality of shielding material as the rubber is viscous, poorly mixed and highly susceptible to an uneven dispersion. Szajerski *et al.* [6] investigated the addition of heavy metal elements to natural rubber in the Bi-W-Gd-Sb system optimized for elastic shielding materials at CT energies and found that the addition of small amounts of rare earth Gd and Sb have more great shielding performance than to pure Bi and W at energies of 55-70 keV.

The main factor determining the ability of a substance to absorb the energy of a ray by the "photoelectric effect" is not the size of the atomic number of the absorbing substance, but the position of the K-layer absorption edge of the absorbing substance (the minimum energy required for a ray particle to knock the K-layer electrons out of the nucleus), i.e. the electron density. The following oxides containing shielding elements (W, Bi, rare earth) were selected for simulation from the viewpoint of shielding performance, economic cost, and environmental protection. The combination was optimized to overcome the weak absorption region existing in each element. Besides, this paper also focuses on a popular and widely used rubber which is SEBS, or styreneethylene-butylene-styrene. This is a thermoplastic elastomer (TPE) that acts like rubber without becoming vulcanized. SEBS is a solid and durable material that is resistant to heat and UV light and is simple to process. It is made by partially and selectively hydrogenating styrene-butadiene-styrene copolymer (SBS), which enhances thermal resilience, weather-ability, and oil tolerance while also making SEBS steam sterile. The objective of the research is to simulate the gamma radiation shielding performance for RTV/SEBS dopant with functional selected nanoparticles using the WinXCom software.

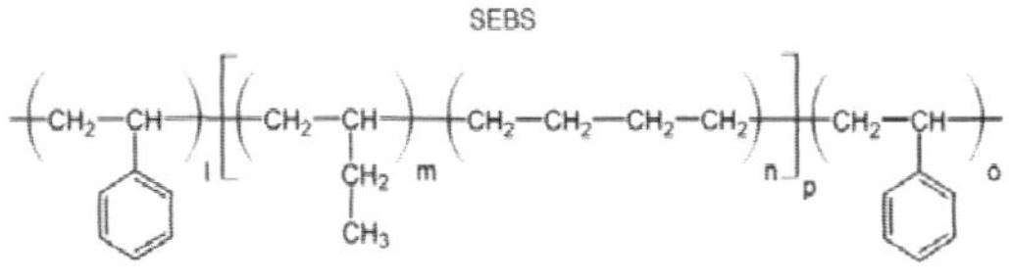

Figure 1: Atomic Structure of SEBS.

2. Methods and materials

The simulation method was done using WinXCOM and with the help of PhyX as a reference.

2.1 Simulation of nuclear radiation shielding performance of functional particles
In the process of simulation using WinXCom software, the atomic number or the symbol which represents the functional nanoparticle will be inserted into the program and followed by entering the additional energies by hand. The output of the unit of all quantities in barn per atom and insert the radiation energy source of a range of 0.001-10000 MeV. The selection of functional elements to simulate using WinXCom are lead (Pb), tungsten (W), bismuth (Bi), gadolinium (Gd), samarium (Sm), and boron (B). Figure 2 shows the overall simulation process using the WinXCom program.

2.2 Simulation of functional particle-modified silicone rubber flexible materials and TPE (SEBS) The process of simulation of functional particle-modified silicone rubber flexible material and TPE is similar to the process of simulating the element used. The name of the compound such as the TPE and silicone rubber will be first selected, together with the name of the functional nanoparticle element to the mixture option, and adjust the percentage of the compound followed by the list of different percentages to get the optimal shielding performance. Figure 2 shows the overall simulation process of the selected compound and mixture.

After the simulation was done, the data was collected in a text file and extracted using Python programming to generate the graph for further analysis.

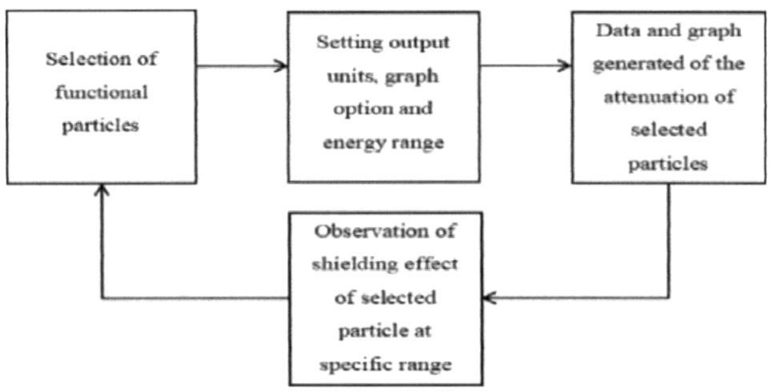

Figure 2. Flow Chart of Particle Simulation Process .

3. Results and discussion

3.1 Simulation of nuclear radiation shielding performance of functional particles
Conventional shielding materials mainly contain the heavy metal element Pb, but Pb has a weak absorption zone during the simulation and is accompanied by strong radiation, especially in the energy 30-150 KeV used in hospital radiotherapy, the weak absorption zone of Pb 40-88 KeV is in it, and the mechanical properties of metallic lead materials are poor, and the specific gravity of lead is large. The design of new alloy composites has to meet the performance requirements. Figure 3 shows the simulation of elements in WinXCOM at photon irradiation shielding performance between 0.001-2 MeV and compiled using Python programming. Every selected element have a weak absorption zone in the energy used in the hospital (30-150 keV) and this is tabulated in Table 1 below.

IOP Conf. Series: Materials Science and Engineering 1285 (2023) 012010 doi:10.1088/1757-899X/1285/1/012010

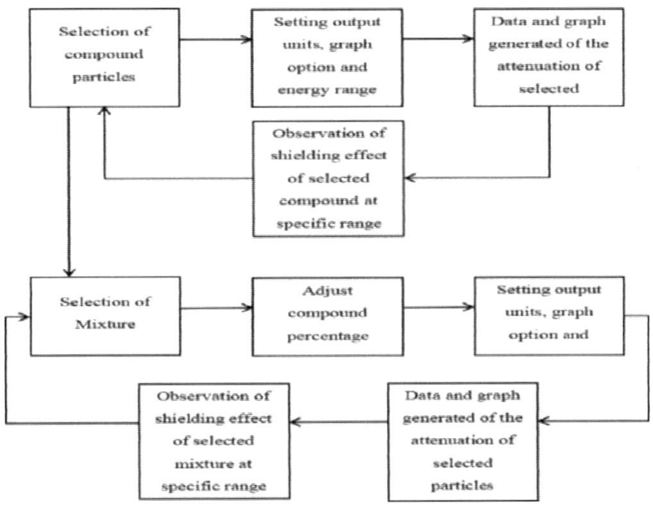

Figure 3. Flow Chart of Compound and Mixture Simulation Process

Table 1: Weak absorption region of the selected elements in the energy region of 30 to 150 KeV .

Element	Weak absorption zone (KeV)
Cd	>50
Sm	40-47
Gd	40-50
Bi	50-91
Sn	>60
Pb	40-88
W	40-69.5
B	Not suitable to do as a shielding material

To ease the viewing, only 5 selected elements were shown in the graph, while the rest was summarized in Table 1. From Table 1, Bismuth, lead and tungsten shows a similar trend in the weak absorption region and this is taken for further research. Based on the common shielding metal composites and the abundance of resources in Malaysia, the simulated shielding effect of Pb and W as a single element was first compared to the bismuth to the photon irradiation shielding performance at 0.001-2 MeV energy, and the simulation results are shown in Figure 5.

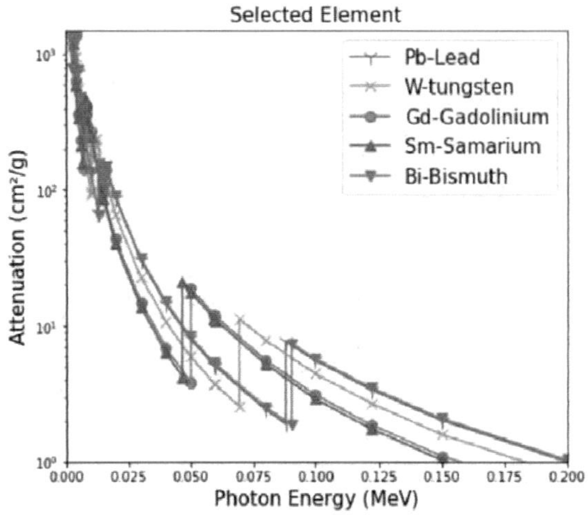

Figure 4. Simulation shielding effect of selected functional particles.

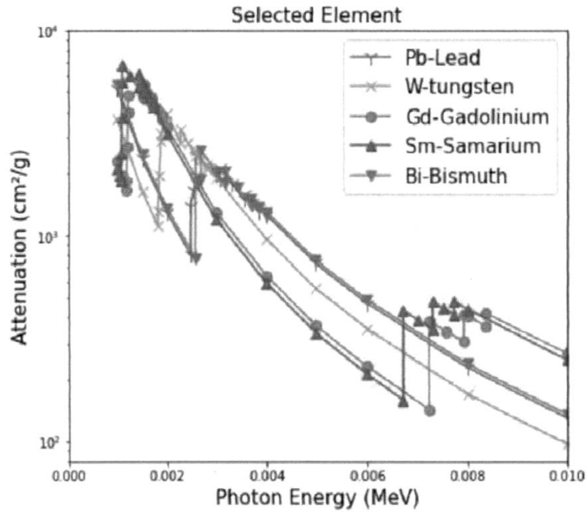

Figure 5. Enlargement view of the Simulation shielding effect of selected functional element (0-0.01 MeV) .

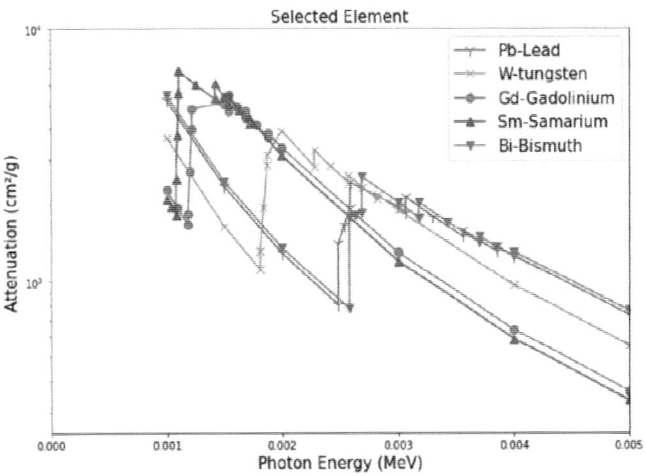

Figure 6. Enlargement view of the Simulation shielding effect of selected functional element (0-0.005 MeV) .

As shown in Figure 7, the relationship between the linear and mass attenuation coefficients, both of which are influenced by the density of the material, and the mass attenuation coefficient. The shielding performance of the particles is closely related to the energy of the rays. Both Pb and W have weak absorption regions in the low energy band, and the weak absorption region of tungsten is narrower at 40 keV-70 keV than the weak absorption region of lead, and the mass attenuation coefficient suddenly increases at slightly greater than the K-layer absorption edge, reaching an increase in the intensity of more than four times. The overall trend is decreased significantly as the energy of the rays increases, with an exponential decreasing pattern of weakening in the low-energy section and a slower linear decreasing pattern in the high-energy section, due to the diverse and complex forms of action at high energies and the reason for this is that there is a variety of complex forms of action at higher energies and the cumulative effect occurs when more electrons are excited over the time. Although the mass attenuation coefficient of Pb is better than that of W, the density of W is greater, so the linear attenuation coefficient of W is better than that of Pb. When the weight of the two materials is not taken into account, the radiation protection performance of a material containing an equal amount of W is better than that of Pb for the same thickness.

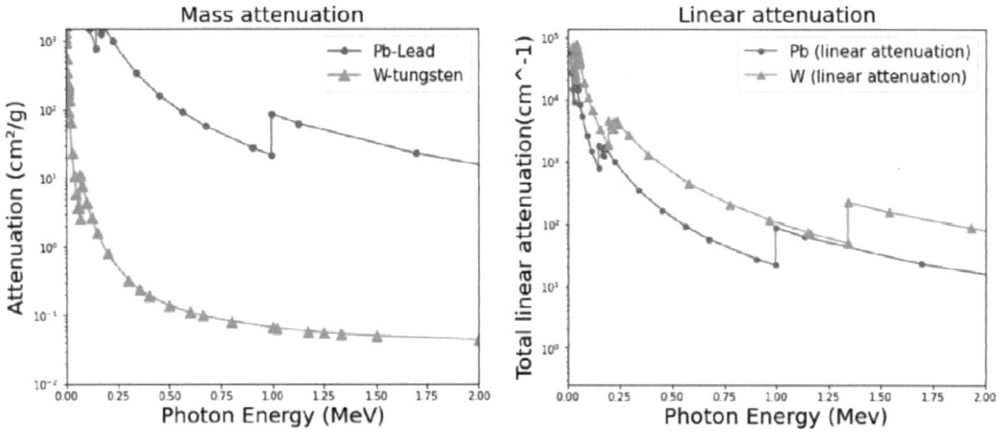

Figure 7: Comparison of the shielding effect between mass attenuation coefficient and subplot linear attenuation (Pb and W)

3.2 Shielding performance of samples

The shielding effect of WO_3 and Gd_2O_3 functional particles in different ratios against several common laboratory sources with different energy gradients and distinctive peaks (^{241}Am: 59.6 keV, ^{57}Co: 122 keV, ^{133}Ba: 356 keV, ^{137}Cs: 662 keV, ^{60}Co: 1173 keV and 1332 keV) was investigated as shown in Figure 8. As can be seen from Figure 8, the shielding performance increases with the increase of Gd_2O_3 content at low energy, and the shielding performance is better when it contains more WO_3 above 122 keV. The shielding performance of the two particles at high energies differs from each other regardless of the variation of their contents.

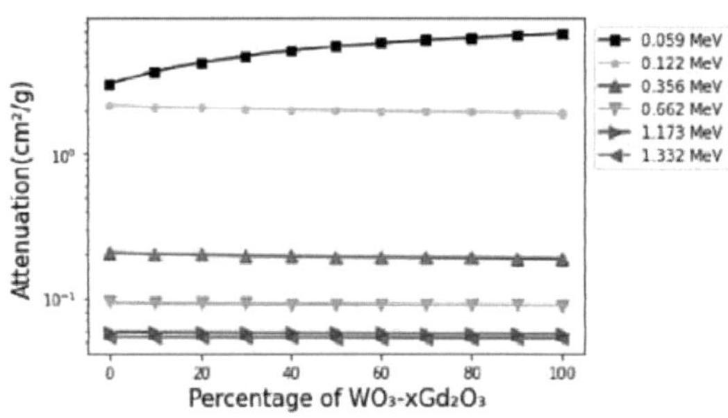

Figure 8. Shielding effect of different ratios of (1-x)WO_3-xGd_2O_3 functional particles (x is the percentage content of Gd_2O_3)

On this basis, a comparison of the shielding performance of the selected more heavy metals and rare earth oxides with the optimized combination of WO_3, photons energy from 0.001 to 2 MeV is shown in Figure 9, as well as the shielding effect for low energy rays from 6 keV to 100 keV. The shielding effect is shown in Figure 3.9 and Figure 3.10. From Figure 3.9, it can be seen that the mass attenuation coefficient of the shielding material decreases as the energy of the radiation increases, and at higher energies, the shielding effect is generally consistent and does not vary much, with Bi_2O_3 being more effective in the mid-energy region.

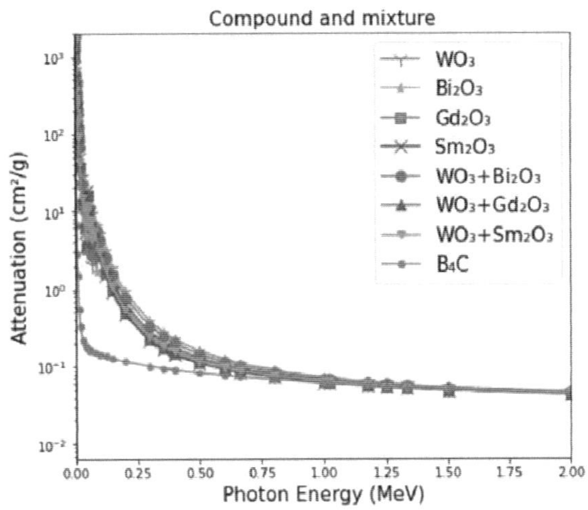

Figure 9. Overall view of optimal combination of functional compounds for shielding effects

However, in the case of common medical radiation at low energies, the differences in the effectiveness of the various substances in low energy radiation protection are more pronounced as shown in Figure 12 and Figure 13. While a single substance shows a single absorption peak, the combined shielding substance shows two strong absorption peaks as shown in Figure 12 and the curve becomes smooth, effectively compensating for the weak absorption region of each substance, especially for the rare earth elements where the K-layer absorption edge is right around this energy, adding a wider energy shielding performance. In other energy regions, however, the shielding is not as effective as for oxides of W and Bi containing heavy atomic coefficients. For special applications in the energy region, this optimized combination of particles can be tailored to maximize material savings and improve shielding performance. The diagram also shows that the shielding performance of the WO_3-based mixture of rare-earth elements increases more at low energies than with the heavier element Bi_2O_3 and that the closer the K-layer absorption edge is to WO_3, the better the shielding increase between the different rare-earth oxides.

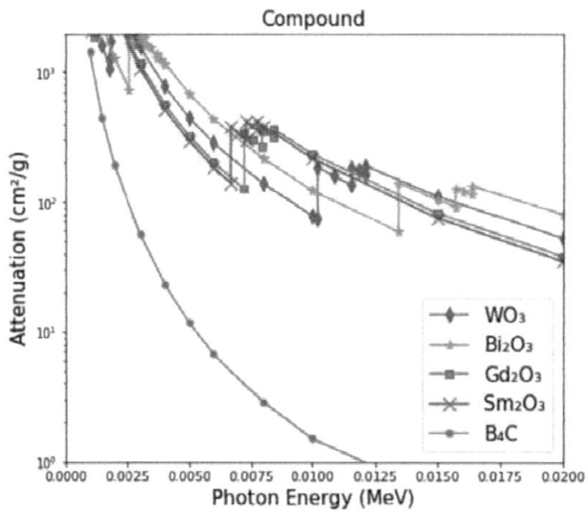

Figure 10. Functional compounds for shielding effects for energy 0-0.02 MeV .

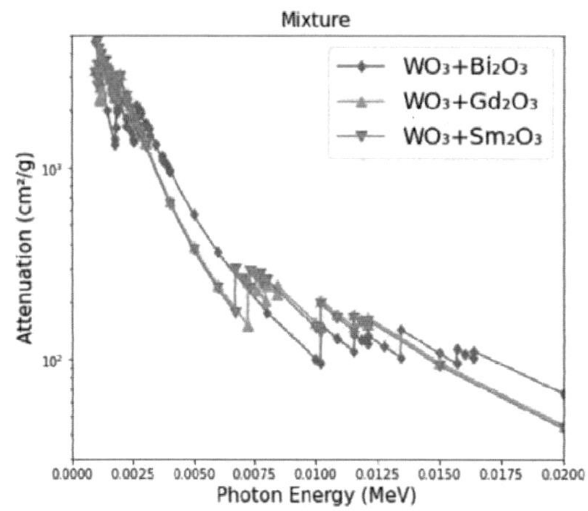

Figure 11. Combination of functional compounds for shielding effects for energy 0-0.02 MeV.

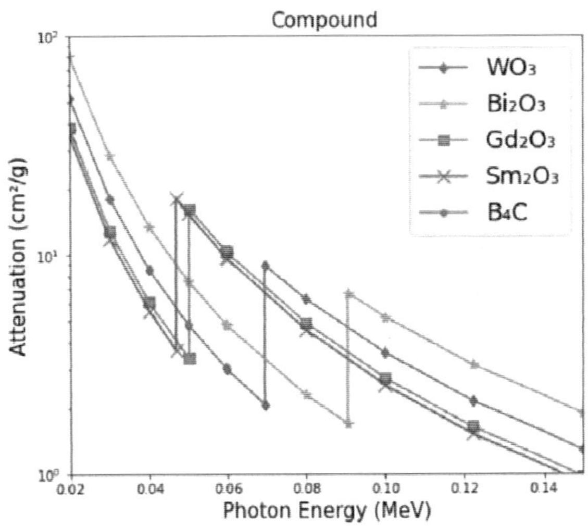

Figure 12. Functional compounds for low-energy radiation shielding (0.02-0.15 MeV)

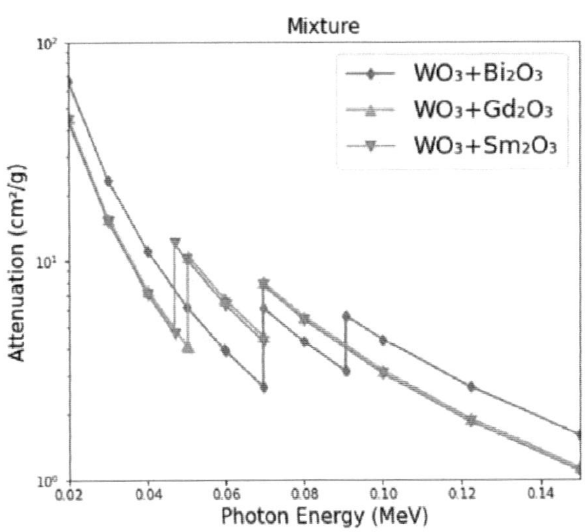

Figure 13. Combination of functional compounds for low energy radiation shielding (0.02-0.15 MeV)

Based on the observations, WO_3+Sm_2O_3+RTV will be the best shielding performance against the gamma radiation. To compare with the previous research paper that being published, the shielding performance of WO_3+Sm_2O_3+RTV has the best performance across the energy range from 0 MeV to 0.015 MeV among the samples in Figure 14.

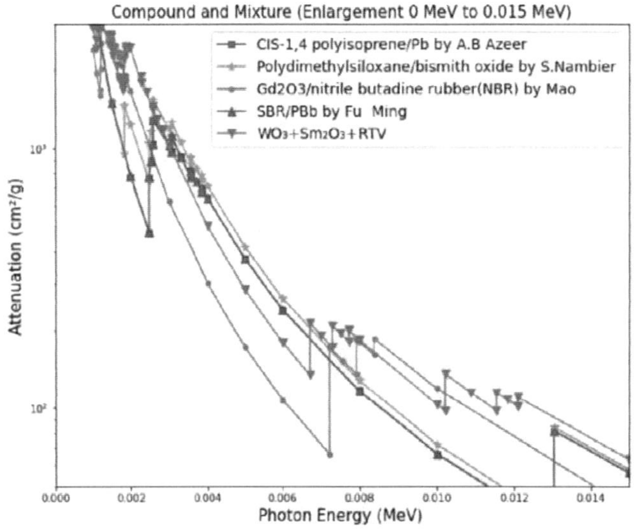

Figure 14. Comparison of WO_3+Sm_2O_3+RTV with previous study for low energy radiation shielding (0.02-0.15 MeV)

4. Conclusion

The main work and result are according to the basic principles of radiation protection and the mechanism of interaction between radiation and matter, several common materials containing shielding protective elements are simulated and calculated using WinXCOM software, including single heavy metals (Pb, W), metal oxides (WO_3, Bi_2O_3, Gd_2O_3, Sm_2O_3), different ratios of functional particles shielding properties, designed room temperature silicone rubber, TPE based composite shielding materials. The simulation result shows that for the same phase of matter, the mass attenuation coefficient is closely related to the energy of the rays, the mass attenuation coefficient decreases significantly with increasing energy of the rays, and in the low energy region, the attenuation is exponential and there is a weak absorption zone, while at high energies, there is a linear attenuation pattern. Using the principle of complementarity of the K-edge absorption region, the substances of multiple combinations of shielding elements are optimally combined, and the shielding performance of the combined material is weaker than that of a single-phase, and the closer the energy of the K-edge absorption region of the two substances, the more obvious the complementary effect of the absorption weak region. The Sm and Gd showed a high shielding performance with the least weak absorption region gap among the selected elements.

For simulation of the functional compound, Sm_2O_3 gave the most pronounced shielding performance peak at 46.8 keV among Gd_2O_3, Bi_2O_3, and WO_3 but the shielding performance dropped significantly when energy increases. When WO_3 is added in, the overall shielding performance increase and the best shielding performance is the combination of Sm_2O_3 and WO_3.

For RTV rubber and SEBS, in the low energy region, the shielding effect increases rapidly with the increase of the doping content of the functional compound, and the shielding effect increases at a slower rate as the doping content exceeds 50phr. In the high-energy region, the shielding effect is relatively gentle. In conclusion, the optimum shielding protection can be achieved if we use the suitable ratio of doping content of functional shielding compounds such as Gd_2O_3 and Sm_2O_3, but if we need the greatest protection, the result suggested that the ratio of doping content of functional compounds to RTV or TPE (SEBS) should be 1:1.

5. References

[1] Azeez A B, Mohammed KS, Zulkepli N N *et al.* 2014 Design of flexible green anti-radiation shielding material against gamma-ray *Materiale Plastice* **51**(3) 300-308.
[2] Mao Y Y, Zhi X, Hu S *et al.* 2015 Preparation of Gd2O3 nano-flakes and fabrication/evaluation of their X-ray shielding rubber nanocomposites with improved mechanical properties *J. of Composite Mater.* **49**(16) 1989-1994.
[3] Nambiar S, Osei E K, Yeow J T W 2013 Polymer nanocomposite-based shielding against diagnostic Xrays *J. of Appl.Polym. Sci.* **127**(6) 4939-4946.
[4] Yue K, Luo W Y, Dong X Q, *et al. 2009* A new lead-free radiation shielding material for radiotherapy *Radiat. Protec. Dosim.* **133**(4) 256-260
[5] Fu M, Wang Y, Li F 2013 Study on the preparation process of SBR rubber-based flexible shielding material *Nuclear Power Process* **34**(6) 165-168.
[6] Szajerski P, Zaborski M, Bem H *et al.* 2014 Optimization of the heavy metal (Bi-W-Gd-Sb) concentrations in the elastomeric shields for computer tomography (CT) *J. of Radioanal. and Nucl. Chem.* **300**(1) 385-391.

Acknowledgement
We would like to thank Universiti Teknologi Malaysia for providing the necessary facility and equipment for the completion of the project.

Preliminary study on radiation shielding properties of sulfur concrete on Mars

Muhammad Nazrif Zamani[1, a)], Muhammad Syahir Sarkawi[2], Asyraf Arif Abu Bakar[3], Sapizah Rahim[3] Mohamad Shazwan Ahmad Shah[1], Sarehati Umar[1], Norhazilan Md Noor[1]

[1]School of Civil Engineering, Faculty of Engineering, Universiti Teknologi Malaysia, Johor, Malaysia
[2]School of Chemical and Energy Engineering, Faculty of Engineering, Universiti Teknologi Malaysia, Johor, Malaysia
[3]Kumpulan Teknologi NDT Termaju (LENDT), Malaysian Nuclear Agency, Bangi, Selangor, Malaysia

[a)]mnazrif2@graduate.utm.my

Abstract. The hypothetical human exploration and inhabitation on Mars would certainly be difficult as the planet itself is vulnerable to the outer space radiation. The outer space radiation consists of the solar particle events (SPEs) that could easily penetrate the very thin atmospheric layers on Mars. Since sulfur concrete is identified as the feasible in situ building material on Mars, its radiation shielding capability is yet to be clarified. Here we report the preliminary radiation shielding properties of the Martian sulfur concrete composed of elemental sulfur and silica with the latter identified on Mars and incorporated by previous related studies upon irradiation with gamma rays generated from the Iridium-192. It is found that the Martian sulfur concrete demonstrated an approximately 50% efficiency upon irradiation. Martian sulfur concrete demonstrated a high potential in radiation shielding capability as well as the structural integrity post irradiation.

1. Introduction

Prior to the 20th Century, the human civilization had witnessed major evolutions with many inventions and enhancements from primitive technologies and further emphasized Mars as the next planetary frontier for human exploration and possible inhabitation as a mean to encourage further technological advancements, ease of human efforts and improved lifestyles [1]. Also known as 'The Red Planet' for its reddish appearance [2], Mars is perceived as the next destination for deeper outer space exploration particularly for its abundant local resources that could be utilized as the materials transportation cost from Earth is very high [3]. For instance, Mars has an abundant source of sulfur that could be extracted to manufacture an in-situ building material. As for the filler, numerous sources highlighted the presence of high amount of non-reactive amorphous silica on Mars [4]–[6]. Previous researchers incorporated amorphous silica as it is a non-reactive filler scattered on Mars that could be integrated together with sulfur to form a solid Martian building material [7], [8]. Therefore, sulfur concrete utilizing silica is perceived as the promising building material on Mars. On the other hand, the very thin atmospheric layers possessed by Mars makes the planet vulnerable to being irradiated by the solar particle events (SPEs) generated from the Sun that are hazardous to human health upon exposure [9]. As a result, the

IOP Conf. Series: Materials Science and Engineering 1285 (2023) 012011 doi:10.1088/1757-899X/1285/1/012011

Martian building material must be able to provide adequate shielding against such irradiation. Recent investigations on the radiation shielding properties have been extensively made on numerous concrete building materials [10]–[15] as illustrated in Table 1. However, the radiation shielding property of a Martian sulfur concrete is yet to be investigated. Therefore, this study aims to investigate the preliminary radiation shielding properties of the Martian sulfur concrete utilizing silica upon exposure to the gamma radiation source from Iridium-192 with the radiation source chosen primarily due to budgetary restriction.

Table 1. Previous related studies on the radiation shielding properties of terrestrial concrete.

Researcher(s)	Year	Radiation source	Type of concrete	Notable ingredient(s)
Roslan et. al [11]	2019	Gamma & neutron	High density concrete	Ferro boron
Kunchariyakun and Sukmak [15]	2020	Gamme & neutron	Conventional concrete	Garnet residue
Suwanmaneechot et. al [10]	2021	Gamma	High density concrete	Metal dust waste
Singh and Singh [12]	2021	Gamma	Green concrete	Fly ash and hematite
Daungwilailuk et. al [13]	2022	Gamma & neutron	Barite concrete	Barite
Zeyad et. al [14]	2022	Gamma	Ultra-high-performance (UHPC) & heavyweight radiation shielding concrete (UHPHSC)	Aggregates (natural sand, magnetite, hematite, ilmenite and barite) + fibers (steel, lead and basalt) + superplasticizer

2. Materials and methods

2.1. Sample preparation

The shape and size of the sulfur concrete specimens utilizing silica illustrated in Figure 2 are fabricated in accordance to BS EN 12390-1:2000. Based on the standard, the basic dimension for the cube dimension should be at least three and a half times the nominal size of the aggregate. Therefore, as the nominal size of the Martian soil simulant used is 1mm, the cube-shaped specimens with dimension 50 mm is used. Such dimension is chosen based on the available. The mix proportion used in this study is 50:50 for silica sand and elemental sulfur respectively [7]. In accordance to the related guideline [16], the elemental sulfur is heated at temperature ranging from 132 to 141°C until it melts prior to gradually adding the silica. The mixture is uniformly stirred and poured into the molds before being left to harden for at least 24 hours. The density of each hardened specimens is recorded followed by the ultrasonic pulse velocity test in accordance to BS EN 12504-4:2004 [17] to determine the porosity of the specimens in terms of longitudinal pulse velocity. The overall flowchart of this study is illustrated in Figure 1.

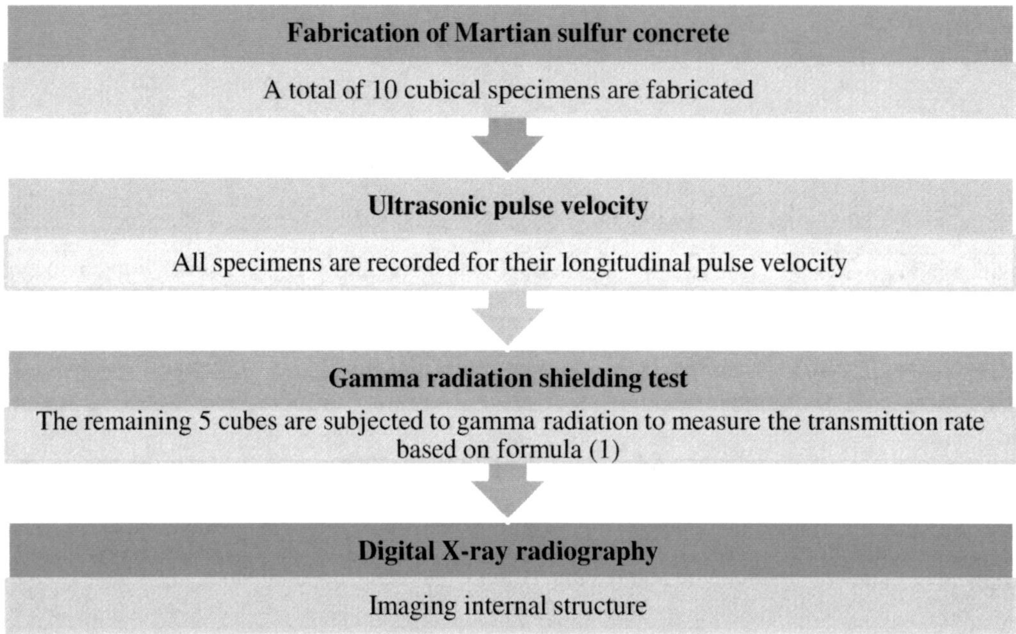

Figure 1. Overall flowchart of this study.

Figure 2. Hardened Martian sulfur concrete.

2.2. Gamma radiation shielding test and digital X-ray radiography

The gamma radiation shielding test is conducted at the Kumpulan Teknologi NDT Termaju (LENDT), Malaysian Nuclear Agency, Bangi, Selangor. The transmission rate is recorded according to the following formula [11]:

IOP Conf. Series: Materials Science and Engineering 1285 (2023) 012011 doi:10.1088/1757-899X/1285/1/012011

$$T = \frac{I}{I_o} \quad (1)$$

where:

$I_o = Initial\ intensity\ (without\ sample)$
$I = Transmitted\ intensity\ (with\ sample)$

The gamma source is consisted of 13.6 mCi Iridium with 0.192 MeV and the gamma scintillation detector is incorporated to measure the gamma intensities. The spacing between the gamma source and the detector is approximately 5 cm shown in Figure 3. Based on Figure 3, lead blocks are surrounding the source, detector and specimen for more accurate readings by confining and restricting the gamma rays during irradiation. In order to visualize the internal structure of the specimens, digital X-ray radiography is used as illustrated in Figure 4.

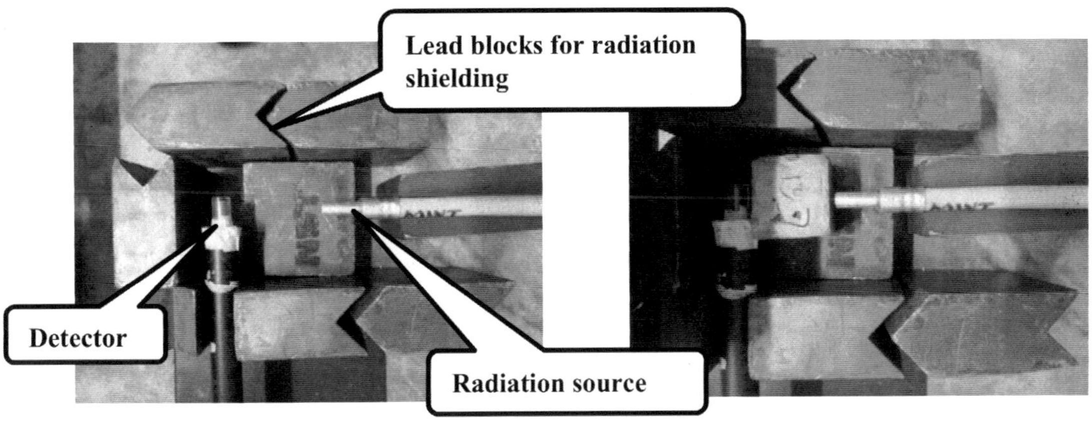

Figure 3. Experimental set up [From left: without sample; with sample

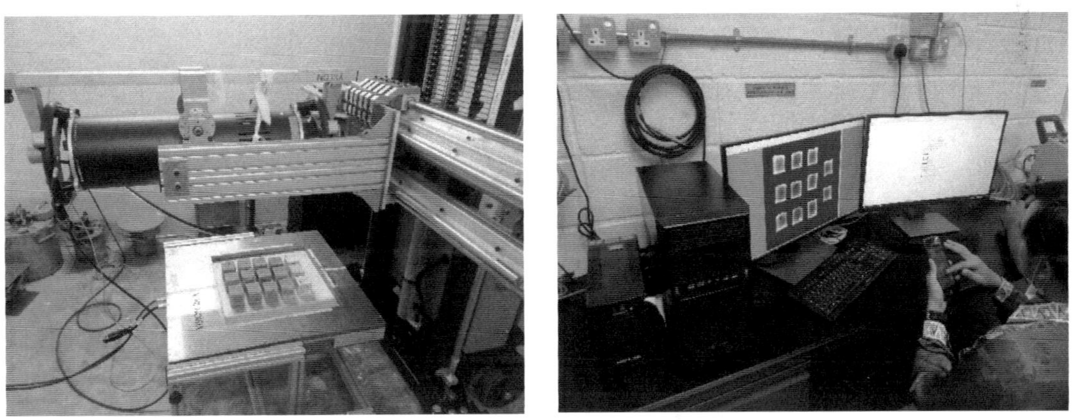

Figure 4. Digital X-ray radiography.

3. Results and discussion

The overall results are illustrated in Table 2. It is found that the Martian sulfur concrete only demonstrated approximately 50% efficiency on radiation shielding properties with average 0.507 transmission rate. According to Wan et. al [7], no chemical reaction occurred between the silica and molten sulfur. Instead, the sulfur only bound the concrete particles together to form a solid sulfur concrete as illustrated in Figure 2. On the contrary, the sulfur concrete's internal structure shown in Figure 5 and 6 also exhibited wide internal voids that resulted from sulfur shrinkage during cooling. According to several sources [7], [16], [18]–[20], molten sulfur initially crystallized to form monoclinic sulfur. Upon cooling, the monoclinic sulfur converted into the denser orthorhombic sulfur. The difference in density triggered internal stress due to the sulfur shrinkage that eventually contributed to potentially wider internal voids based on the morphological images of this study illustrated in Figure 5. Consequently, the penetration of radiation source was higher.

Table 2. Overall results of the concrete specimens.

Specimen code	Density (kg/m^3)	Velocity (mm/μs)	Transmission rate (I/I_o)
G1	2200	3.01	0.506
G2	2200	3.13	0.504
G3	2320	3.05	0.512
G4	2240	3.07	0.506
Average	2240	3.07	0.507

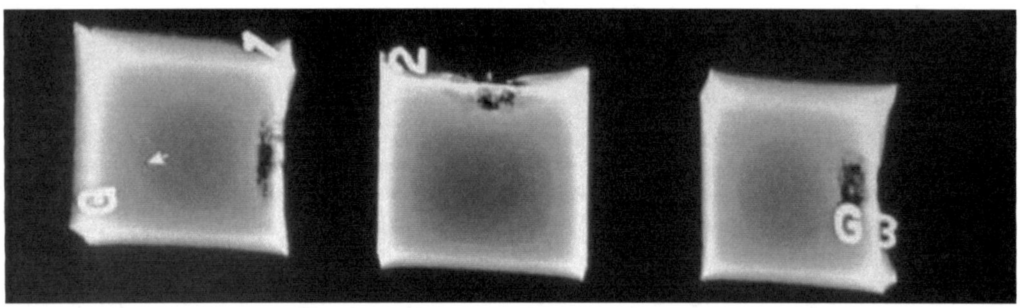

Figure 5. Internal porosity of the concrete specimens from X-ray digital radiography.

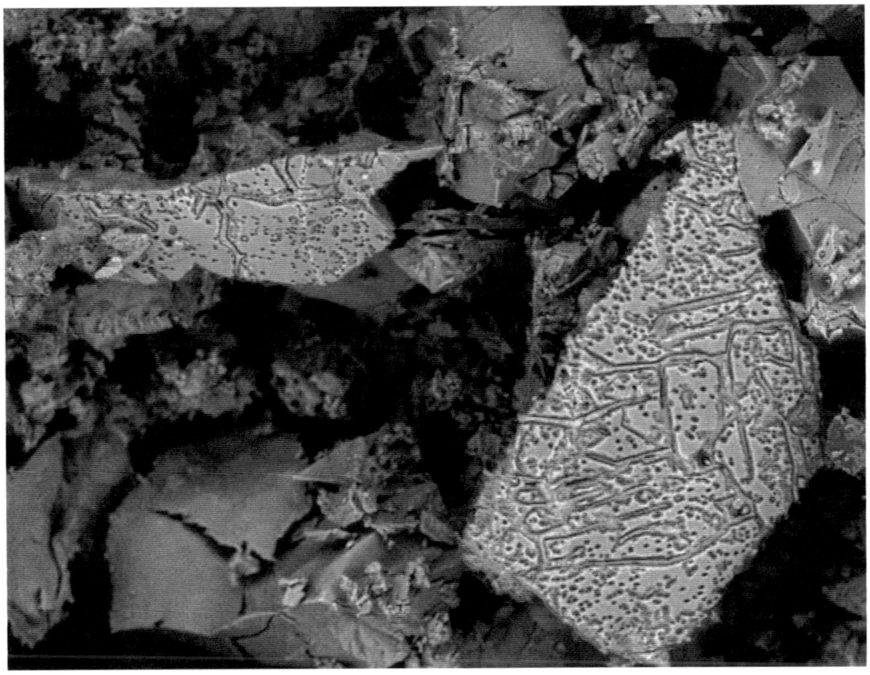

Figure 6. [From left] Internal structure of the Martian sulfur concrete.

4. Conclusion

In order to address the need for radiation shielding on the hypothetical Martian building material due to the inevitable irradiation on Mars, the Martian sulfur concrete comprised of elemental sulfur and silica is fabricated and evaluated in terms of transmission rate and compressive strength before and after irradiation. The radiation shielding property of such concrete addressed in this preliminary study provides a prediction that it could exhibit more optimum radiation shielding properties provided that the findings of this preliminary study motivate future studies in improving and addressing this prediction. Recommended future study would be investigating the influence of varying sulfur and silica composition on the radiation shielding property of the Martian sulfur concrete. The utilization of the Martian soil simulant developed by researchers that almost match the actual Martian soil [21] is also highly recommended as it provides clearer predictions on the radiation shielding effectiveness of the Martian sulfur concrete.

5. Reference

[1] Zubrin R 2000. The Significance of the Martian Frontier *Case Mars VI Mak. Mars an Afford. Destin.* 27–37,
[2] Ezell E C and Ezell L N 1984 *On Mars: Exploration of the Red Planet, 1958-1978* **4212**. (Scientific and Technical Information Branch, National Aeronautics and Spac)..
[3] Davis G, Montes C, and Eklund S 2017 Preparation of lunar regolith based geopolymer cement under heat and vacuum *Adv. Sp. Res.* **59**(7) 1872–1885 doi: 10.1016/j.asr.2017.01.024.
[4] Fenton L K, Gullikson A L, Hayward R K, Charles H and Titus T N 2018 The Mars Global Digital Dune Database (MGD 3): Global patterns of mineral composition and bedform stability," *Icarus* **330** 189–203 doi: 10.1016/j.icarus.2019.04.025.
[5] Rice M. S. *et al.* 2013 Reflectance spectra diversity of silica-rich materials: Sensitivity to environment and implications for detections on Mars *Icarus* **223**(1) 499–533 doi:

10.1016/j.icarus.2012.09.021.
[6] A. N. Scott and C. Oze, "Constructing Mars: Concrete and Energy Production From Serpentinization Products," *Earth Sp. Sci.*, vol. 5, no. 8, pp. 364–370, 2018, doi: 10.1029/2017EA000353.
[7] Wan L, Wendner R and Cusatis G 2016 A novel material for in situ construction on Mars: experiments and numerical simulations *Constr. Build. Mater.* **120** 222–231 doi: 10.1016/j.conbuildmat.2016.05.046.
[8] Li H *et al.* 2021 Development of a novel material and casting method for in situ construction on Mars *Powder Technol.* **390** 219–229 doi: 10.1016/j.powtec.2021.05.054.
[9] Meurisse A, Cazzaniga C, Frost C, Barnes A, Makaya A, and Sperl M 2020 Neutron radiation shielding with sintered lunar regolith *Radiat. Meas.* **132** doi: 10.1016/j.radmeas.2020.106247.
[10] Suwanmaneechot P, Bongkarn T, Joyklad P and Julphunthong P2021 Experimental and numerical evaluation of gamma-ray attenuation characteristics of concrete containing high-density materials *Constr. Build. Mater.* **294** doi: 10.1016/j.conbuildmat.2021.123614.
[11] Roslan M K A, Ismail M, Kueh A B H and Zin M R 2019M High-density concrete: Exploring Ferro boron effects in neutron and gamma radiation shielding," *Constr. Build. Mater.* 215 718–725 doi: 10.1016/j.conbuildmat.2019.04.105.
[12] Singh S and Singh K 2021On the use of green concrete composite as a nuclear radiation shielding material *Prog. Nucl. Ener.***136** doi: 10.1016/j.pnucene.2021.103730.
[13] Daungwilailuk T, Yenchai C, Rungjaroenkiti W, Pheinsusom P, Panwisawas C, and Pansuk W, 2022 Use of barite concrete for radiation shielding against gamma-rays and neutrons *Constr. Build. Mater.* **326** oi: 10.1016/j.conbuildmat.2022.126838.
[14] Zeyad A M, Hakeem I Y, Amin M, Tayeh B A and Agwa I S 2022 Effect of aggregate and fibre types on ultra-high-performance concrete designed for radiation shielding," *J. Build. Eng* **58** doi: 10.1016/j.jobe.2022.104960.
[15] Kunchariyakun K. and Sukmak P 2020 Utilization of garnet residue in radiation shielding cement mortar *Constr. Build. Mater.* **262** doi: 10.1016/j.conbuildmat.2020.120122.
[16] Fontana J J *et al.* 1998 *Guide for Mixing and Placing Sulfur Concrete in Construction* **93**1–12,.
[17] BSI Standards 2004 Testing concrete — Part 4: Determination of ultrasonic pulse velocity *Br. Stand.* **3** 18.
[18] Fediuk R. *et al.* 2020 A critical review on the properties and applications of sulfur-based concrete *Materials* (Basel) **13**(21) 1–23 doi: 10.3390/ma13214712.
[19] Dugarte M, Martinez-Arguelles G and Torres J 2019 Experimental evaluation of modified sulfur concrete for achieving sustainability in industry applications *Sustain.* **11**(1) 1–16 doi: 10.3390/su11010070.
[20] Mohamed A M O and El-Gamal M 2010 *Sulfur concrete for the construction industry: a sustainable development approach* (J. Ross Publishing).
[21] Liu J. *et al.* 2022 In-situ resources for infrastructure construction on Mars: A review *Int. J. Transp. Sci. Technol.* **11**(1) 1–16 doi: 10.1016/j.ijtst.2021.02.001.

Acknowledgement

The authors are very grateful to the Kumpulan Teknologi NDT Termaju (LENDT), Malaysian Nuclear Agency, Bangi, Selangor for the support and guidance in this research under the grant number 08G82 and UTM ER, Q. J130000.3851.20J64.

IOP Conf. Series: Materials Science and Engineering 1285 (2023) 012012 doi:10.1088/1757-899X/1285/1/012012

Attenuation range of particle radiations (alpha and beta) in human skin by Monte Carlo simulation

Nor Alia Aqilah Nawawi[1], Suffian Mohamad Tajudin[1,a)], Fairuz Mohd Nasir[1], Nurul Syazwina Mohamed[1], Nazifah Abdullah[1], Nur Hamizah Mohd Zainudin[1] and Juliana Mohd Radzi[1]

[1]Faculty of Health Sciences, Universiti Sultan Zainal Abidin (UniSZA), Terengganu, Malaysia.

[a]suffian@unisza.edu.my

Abstract. Particle radiations such as alpha and beta particles are more hazardous to humans than photons because of their higher linear energy transfer (LET). It is important to analyze the profile of particle radiations when interact with human skin. It is generally said that α particles can be stopped by a piece of paper while β rays can be stopped by an aluminum plate. A Monte Carlo simulation (PHITS code) is used for determining the attenuation and range of commonly available alpha (Po-210, Po-214, Am-241) and beta sources (Y-90) in human skin. These particle radiations were set at the surface of several materials such as skin, paper, water, aluminium and air to see their distribution within the materials. The PHITS code outputs were verified by comparing it with other literature for its range in air. Results showed that alpha particles barely pass through the outer layer of dead human skin and definitely can be stopped by a piece of paper. From the particle trajectory output, fluence of α particles from Po-210, Po-214, and Am-241 are completely stopped at 0.005 cm, 0.009 cm, and 0.012 cm in a paper, respectively, with no secondary particles are generated. On the other hand, beta particles would easily penetrate human skin and its range in human skin is ~0.8 cm for Y-90, while it would stop at ~0.3 cm in aluminium. In conclusion, based on these simulations, it was found that this study had justified the common sense on the shielding for the particle by using paper for alpha and aluminium for beta particle as a function of particle energy. Calculation with PHITS can be used for range calculation of various radiation types, including neutron and proton particles.

1. Introduction

In workplaces with radioactive sources and radiation-generating devices such as accelerators and X-ray tubes, radiation dose must be appropriately controlled to prevent hazards caused by exposure to radiation [1 – 4]. Particle radiation, especially alpha is the most dangerous if it is inside the body after being ingested, swallowed or inhaled. This internal exposure of particle radiation is harmful to human health because it can easily absorb by cells while for electromagnetic radiation or photon, they are not as dangerous as alpha because they usually pass through the cell and less likely to be absorbed by the cell [5]. In the case of external exposure, alpha radiation is not dangerous as it cannot penetrate the outer layer of dead skin cells and can be stopped with a piece of paper. At the same time, beta radiation can penetrate skin a few centimeters which are also a sign of external health risk. However, the threat is still primarily from internal exposure from inhaled or ingested material [6]. Both of them can easily be shielded if kept outside of the body. The interaction of ionizing radiation (alpha and beta) whether internal contamination or external exposure with the human body, leads to biological effects that can appear as clinical signs and symptoms.

Charged particles lose its energy by ionization process when passing through matter until their energy becomes zero. The range of particles is the maximum distance the particle travels from its source through medium or matter. The range depends on particle properties such as the initial energy and its characteristics. The mean range of particle radiations can be estimated by integrating the inverse of stopping power over energy. Figure 1 shows the mean range of alpha and beta radiations. The range of charged particles varies within the thickness of material [7], where the beta range is longer than alpha in the same material.

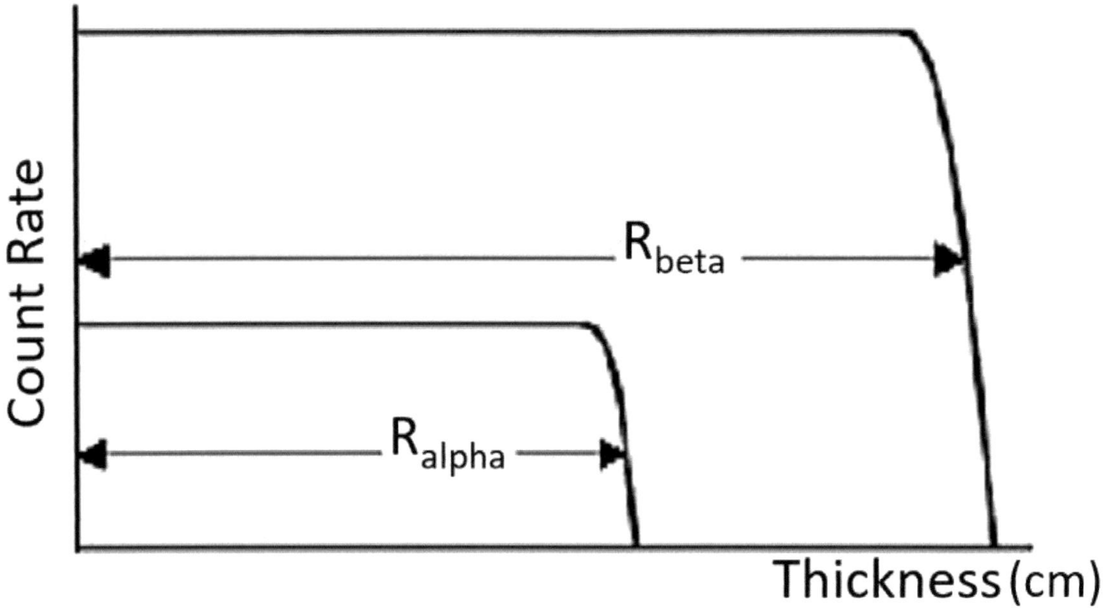

Figure 1. The mean range of the particle radiations [7].

According to United States Nuclear Regulatory Commission (U.S. NRC, 2011 [8]), the beta range could be expressed as a density thickness that has units of g/cm^2. Several equations can estimate the range of betas to design the thickness of the shield. Feather's Rule can be applied to beta particles with maximum energy above 0.6 MeV. The equation is shown below;

$$Range\left(\frac{g}{cm^2}\right) = 0.542(E) - 0.133 \qquad (1)$$

where E is the maximum energy of beta particles in MeV.

In this study, the attenuation range of particle radiations of several materials were evaluated by Monte Carlo simulation (PHITS code [9, 10]). In the first case, the attenuation range of alpha particle was calculated in the air for verification purposes before subsequent calculations for human skin and paper materials. In the next case, we include the calculation of attenuation range in human skin and aluminium for beta particle source. Three alpha particle sources of higher energy such as Po-210, Po-214 and Am-241 were employed, including one beta source of Y-90 that is commonly available. The branching ratio of the particle sources were sampled as the JRIA data book [11] as shown in Figure 2 (a - d) for the output of PHITS code for sources check. From the Figure 2 (a) and (b), the alpha sources of Po-210 (5.304 MeV; 100 %) and Po-214 (7.687 MeV; 100 %) are monoenergetic where they will travel the same distance in materials before being absorbed. While Am-241 (Figure 2 (c)) emitted alpha particle at 5.486 MeV (84.7 %), but it also produces energy alpha of 5.388 MeV (1.6 %) and 5.443 MeV (13 %) at the same time. For Y-90, the electron spectrum from the beta decays is a continuous spectrum with the maximum energy (E_{max}) at 2.28 MeV, with average beta energy of 0.934 MeV, as shown in Figure 2 (d).

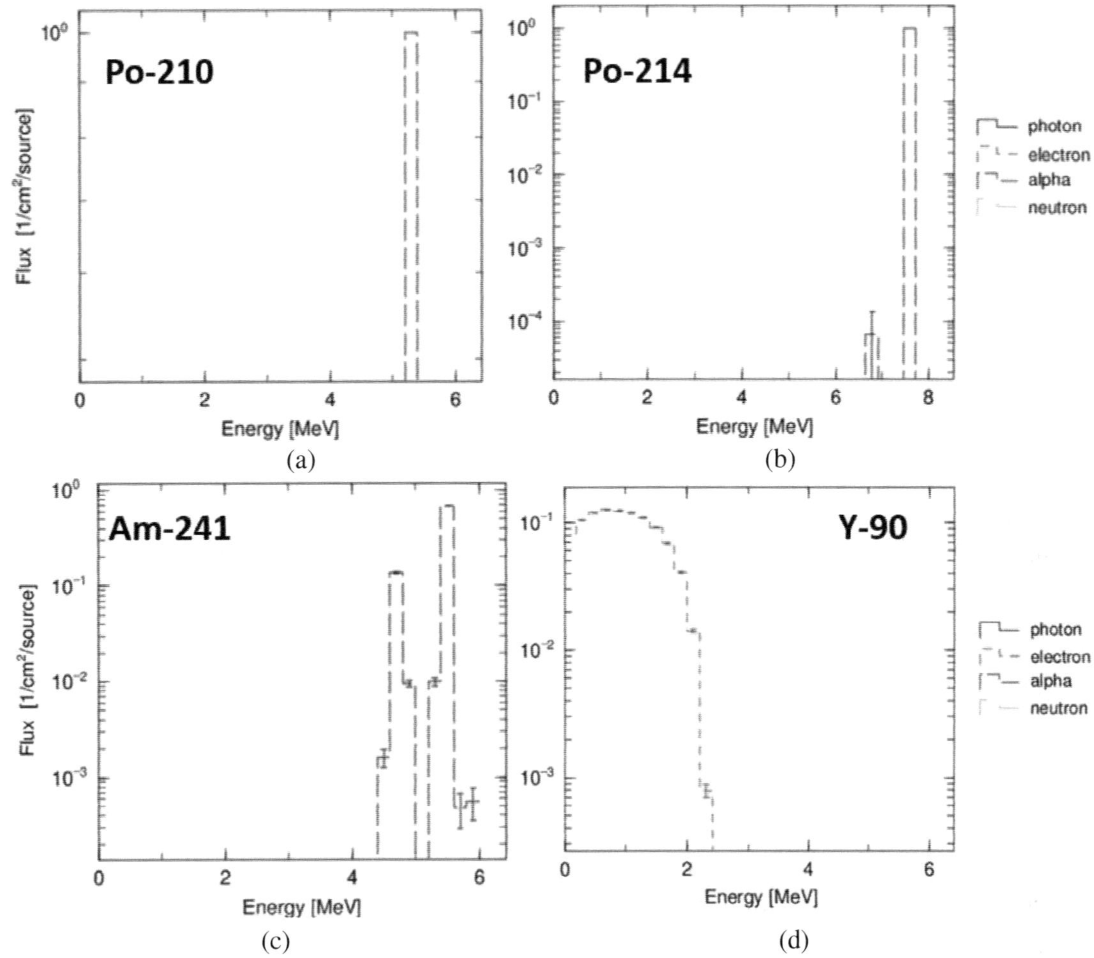

Figure 2. Particle radiation energy spectra as a source input in the calculation. The main energy of alpha particles emitted from Po-210 (a), Po-214 (b) and Am-241 (c) are 5.30 MeV, 7.69 MeV and 5.49 MeV, respectively. For Y-90 (d), continuous spectra with E_{max} is at 2.28 MeV [11].

2. Calculated attenuation range of alpha sources

PHITS code has been developed at the High Energy Accelerator Research Organization, Japan. The code is a general purpose package for the Monte Carlo simulation to transport particles and photons in an arbitrary geometry with energies above a few keV up to several hundred GeV. PHITS is useful for particle transport simulation in a mixed radiation field owing to its applicability to various radiation types. To verify our Monte Carlo code, we calculated the range of Po-210 (5 MeV) and Po-214 (7.69 MeV) in the air. The density and atomic compositions of air are based on the National Institute of Standards and Technology (NIST) reference database [12].

Table 1. The density of materials used for calculation is based on the NIST database [12].

Material	Density (g cm^{-3})
Skin (ICRP)	1.10
Paper	0.82
Air	0.0012
Aluminium	2.70

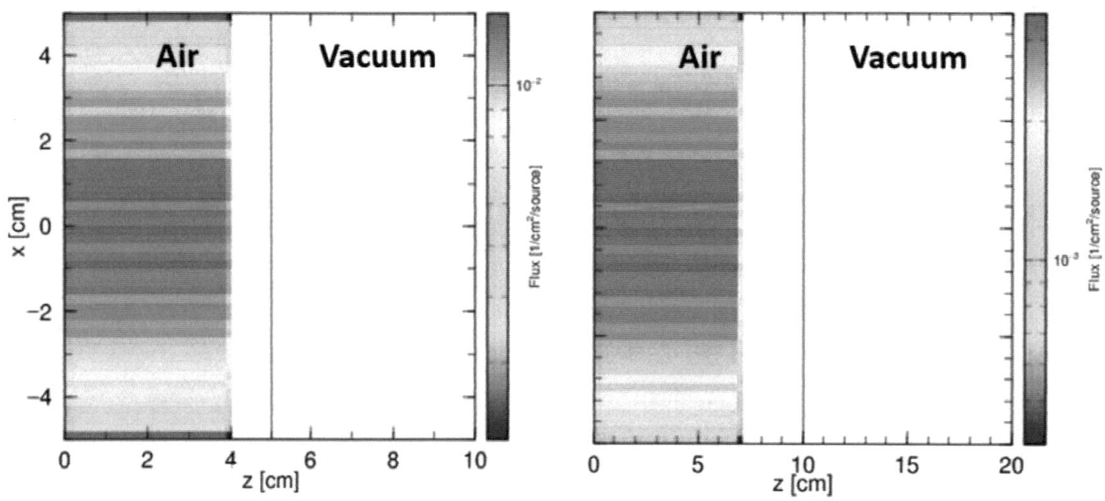

(a) (b)

Figure 3.: The alpha particle range in air emitted by Po-210 (a) and Po-214 using PHITS code stopped at ~4 cm and ~7 cm, respectively.

Figure 3 shows the range of alpha particles in air emitted by Po-210 (a) and Po-214 (b) using the PHITS code. All the alpha particles are almost entirely stopped in the air at a distance of 4 cm and 7 cm, respectively. The outputs are the same as those of Nikezic and Yu [13] using another Monte Carlo software (SRIM code). The calculation were then extended for other materials such as human skin (ICRP), paper and aluminium, as in Table 1. The elemental compositions are obtained from the NIST database.

Table 2 shows the range of alpha particle sources in human skin and paper. On average, the alpha particles range of penetration is 34 % higher in paper than in human skin due to the higher density of human skin than paper. For skin, the outermost epidermal layer is called stratum corneum. It consists of dead cells and is the primary barrier to chemical transfer through the human skin [14]. In addition, the stratum corneum is the barrier to protect underlying tissue from dehydration or infection. The thickness of the stratum corneum is between 10 and 40 micrometers which are about 0.001 to 0.004 cm. For example, the range of Po-210 in the skin is 0.004 cm and therefore would barely pass through the skin, meanwhile, for Po-214 and Am-241 are 0.0072 cm and 0.0082 cm, respectively. Nevertheless, alpha particles can easily be shielded by a single sheet of paper, which needs less than 0.1 mm thickness to attenuate all alpha sources studied.

Table 2. The attenuation range of three alpha particle sources in human skin and paper.

Materials	Range of alpha particles (in cm)		
	Po-210	Po-214	Am-241
Skin	0.004	0.0072	0.0082
Paper	0.0052	0.009	0.012

3. Calculated attenuation range of beta source

The beta source (Y-90) calculation was performed to see the impact of different types of particle on the same human skin, including an aluminium commonly used as its attenuator [15]. Figure 4 shows the Y-90 beta range in 1 cm of human skin (a) and aluminium (b), respectively.

It is clear that beta particles can be fully stopped by ~0.4 cm of aluminium, but could easily penetrate human skin. Approximately 1 cm of human skin thickness is needed to completely stop the beta range from the Yttrium-90 source. Kim et al. [16] had mentioned in their article that the average penetration depth in human tissue of Y-90 is about 0.24 cm. In such a case, a thin of aluminium (~0.4 cm) is adequate to stop beta particles entirely.

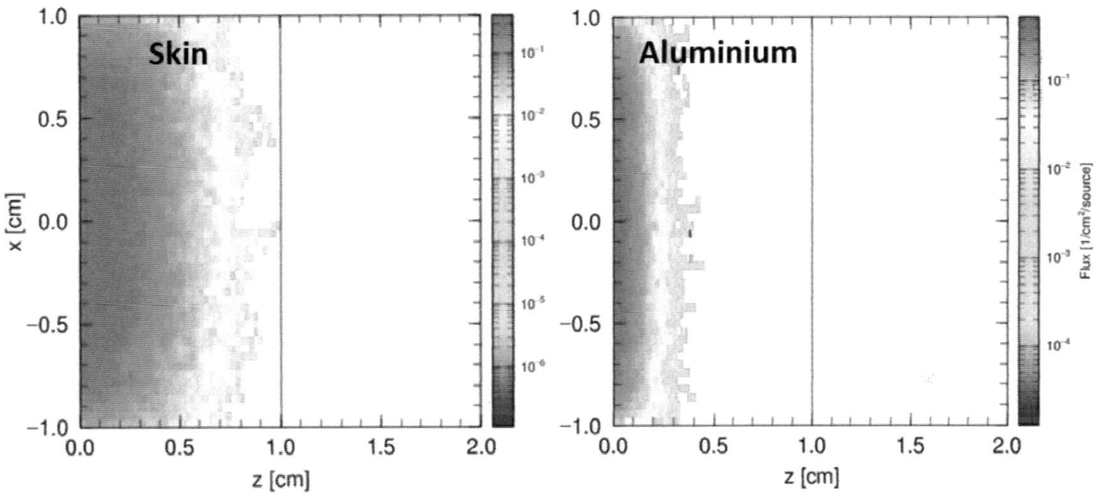

Figure 4. The Y-90 beta range in 1 cm of human skin (a) and aluminium (b), respectively.

The ease of penetrating through the human body depends on the types of radiation. Comparing these two types of particle radiations, alpha particles have the greatest mass. Due to their heavy mass, they have the greatest ability to damage tissue and the highest ionizing power but have the least penetration of matter, especially human skin. They can be stopped by a piece of paper and barely pass through the outer layer of the dead skin of human body. However, alpha particles have high LET which means that most of energetic particles deposited closely together so less space between these generated particles can make them release all their energy in a few cells and will cause the DNA damages is more likely, so high LET particles can be identified as one of the major contributor to health risk. These high LET particles deposit more energy over a short distance may be the reason for a dramatic event that causes damage to DNA that will affect the human organs. Even with the same radiation dose, an alpha particle will cause more damage compared to low LET such as X-rays because photons deposit their energy gradually over a long distance thus, so they are less damaging than alpha [17]. The external radiation hazards of alpha to human skin are not a major concern as they barely have enough energy to penetrate the outer layer of skin. However, it remains a threat if internal contamination happens as it can damage sensitive living tissue. As an example, a proton-targeted therapy for cancer treatment could induce an internal contamination that would generate more secondary particles (particularly the alpha particles) which potentially induce extra dose to tumor region [18].

On the other hand, beta particles are more penetrating than alpha particles. They also can travel further due to their smaller mass. However, they are less damaging to the DNA and living tissues as they have low LET compared to alpha particles and the ionization that they produce also more widely spread. Beta particles can pass through and penetrate a few millimeters of human skin but can be fully stopped by a thin aluminium sheet. Since they can penetrate a tiny part of human skin, they can cause some damage like skin burns when the doses are very high. But, it still does not pose significant damage and beta particles are hazardous and can do internal damage to cells and organs if swallowed or inhaled, like alpha particles [19].

4. Conclusion

The types of radiation vary in their ability to penetrate the material. Alpha particles are the least penetrating compared to beta particles even though the alpha energies are higher than beta particles due to their heavy mass. Alpha particles barely pass through the outer layer of the dead skin of human body and can be stopped by a single sheet of paper that has <0.1 mm thickness to attenuate all alpha sources. On the other hand, beta particles easily penetrate up to 1 cm of human skin thickness and can be entirely shielded with 0.4 cm of aluminium. PHITS simulations confirmed the common sense on the shielding profiles of α and β. Lastly, the Monte Carlo code could be used to design shielding for particle radiations such as alpha, beta, neutron, and proton, including the range of particle radiations in the human body for internal contamination (blood, tissue, bone).

5. References

[1] Aminordin Sabri A H, Abdul Aziz M Z, Olukotun S F, Tabbakh F, Tajudin S M 2020 Study on the shielding materials for low-energy gamma sources *IOP Conf. Ser.: Mater. Sci. and Eng.* **785**(1), 012007 https://doi.org/10.1088/1757-899x/785/1/012007.

[2] Sabri A, Abdul Aziz M Z, Olukotun S F, Tajudin S M 2022 Calculation of attenuation parameter for Ir-192 gamma source in shielding material *J. of Med. Phys.* **47**(1)34–39. https://doi.org/10.4103/jmp.jmp_86_21.

[3] Tajudin S M, Aminordin Sabri A H, Abdul Aziz M Z, Tabbakh F 2021 Gadolinium-doped polymeric as a shielding material for X-ray. *IOP Conf. Ser. Mater. Sci. and Eng.* ***1106***(1) https://doi.org/10.1088/1757-899X/1106/1/012010.

[4] Tajudin S M, Tabbakh F 2019 Biological polymeric shielding design for an X-ray laboratory using Monte Carlo codes. *Radiol. Phys. and Tech.* **12**(3), 299–304 https://doi.org/10.1007/s12194-019-00522-w.

[5] Donya M, Radford M, ElGuindy A, Firmin D, Yacoub M H 2014 Radiation in medicine: Origins, risks and aspirations. *Global Cardiol. Sci. and Prac.* **2014**(4) https://doi.org/10.5339/gcsp.2014.57.

[6] Mirion Technologies 2015 Types of ionizing radiation. https://https://www.mirion.com/learning-center/radiation-safety-basics/types-of-ionizing-radiation.

[7] Ghanbari F 2011 GNEII curriculum (TAMU): Week 2: Introduction to radiation shielding: basic concepts (U) (Texas A&M University) retrieved April 7, 2022: https://www.osti.gov/biblio/1690359.

[8] Nuclear Regulatory Commission (NRC) *2011* Shielding radiation https://www.nrc.gov/docs/ML1122/ML11229A721.pdf.

[9] Sato T, Iwamoto Y, Hashimoto S, Ogawa T, Furuta T, Abe S I, Kai T, Tsai P E, Matsuda N, Iwase H, Shigyo N, Sihver L, Niita K 2018 Features of particle and heavy ion transport code

system (PHITS) version 3.02. *J. of Nucl. Sci. and Tech.* **55**(6) 684–690. https://doi.org/10.1080/00223131.2017.1419890.

[10] Tajudin S M, Sabri A H A 2019 SIMU-RAD programme: a learning tool for radiation (photons and charged particles) interaction *Polish J. of Med. Phy. and Eng.* **25**(3) 189–192 https://doi.org/10.2478/PJMPE-2019-0025.

[11] Japan Radioisotope Association 2011 Radioisotope pocket data book (Tokyo: Maruzen).

[12] Compositions of materials used in STAR databases. (n.d.) retrieved January 8, 2022: https://physics.nist.gov/cgi-bin/Star/compos.pl.

[13] Nikezic D, Yu K N 2016 Alpha-particle fluence in radiobiological experiments *J. of Radiat. Res.* **58**(2) 195–200 https://doi.org/10.1093/jrr/rrw106.

[14] Hopewell J 1990 The skin: its structure and response to ionizing radiation *Int. J. of Radiat. Biol.* **57**(4), 751–773 https://doi.org/10.1080/09553009014550911.

[15] Van Pelt W R, Drzyzga M 2007 Beta radiation shielding with lead and plastic: effect on Bremsstrahlung radiation when switching the shielding order *Health Phys.* **92**(1) S13–S17 https://doi.org/10.1097/01.hp.0000248127.71398.b1.

[16] Kim Y C, Kim Y H, Uhm S H, Seo Y S, Park E K, Oh S Y, Jeong E, Lee S, Choe J G 2010 Radiation safety issues in Y-90 microsphere selective hepatic radioembolization therapy: possible radiation exposure from the patients. *Nucl. Med. and Molecul. Imaging*, **44**(4) 252–260 https://doi.org/10.1007/s13139-010-0047-7.

[17] Range | Particle Radiation 1998 *Encyclo. Britannica* retrieved March 21, 2022: https://www.britannica.com/science/range-particle-radiation.

[18] Tabbakh F, Hosmane N S, Tajudin S., Ghorashi A, Morshedian, N 2022 Using 157Gd doped carbon and 157GdF4 nanoparticles in proton-targeted therapy for effectiveness enhancement and thermal neutron reduction: a simulation study *Scientific Reports*, **12**(1) https://doi.org/10.1038/s41598-022-22429-0.

[19] Environmental Protection Agency 2022 *Environ. Topics* retrieved February 18, 2022: https://www.epa.ie/environmental-topics/#d.en.71466.

IOP Conf. Series: Materials Science and Engineering 1285 (2023) 012013 doi:10.1088/1757-899X/1285/1/012013

Effects of gamma rays on normal-high-density concrete

Farah Ainin Sofiya Muhd Syazli[1] and Mohd Idzat Idris[1,a)]

[1]Nuclear Science Program, Applied Physics Studies, Faculty of Science and Technology, Universiti Kebangsaan Malaysia, Bangi, Selangor, Malaysia

[a]idzat@ukm.edu.my

Abstract. Concrete is one structural and shielding material used against ionising radiation. It is used in radioactive environment such as radiotherapy rooms and radioactive safe keeping. Optimise concrete is evaluated by its thickness and also course and fine aggregate. Mixture of course and fine aggregate in concrete and its grade play an important role as good radiation absorber. However, concrete with unsuitable mixture will display cracks and will cause radiation leakage which could harm the public's health. Therefore, this research is carried out using normal concrete with high density to determine the optimal grade. Four grades of concrete were produced, with mixture of cement, water and course-fine aggregate i.e. limestone and silica sand. Attenuation coefficient is measured using gamma spectrometer NaI (Tl) and single channel analyser (SCA). Radiation source used are Cs-137 and Co-60with photon energies of 0.662 MeV, 1.173 MeV and 1.332 MeV. Calculation of attenuation coefficient for concrete with thickness of 2.6 cm, 5.2 cm, 7.8 cm, 10.4 cm dan 13 cm shows increase in attenuation coefficient (<1.0 cm-1) for each grade, namely M15, M25, M35 dan M45. The result shows optimum concrete grade is M35 as the attenuation coefficient, which is between 0.04 - 0.18 cm-1, is the lowest in comparison with the other grades. Thus normal high density concrete could be used as effective shielding material and viable.

1. Introduction

Concrete has been widely used worldwide in the construction industry because of its cheap price and mechanical properties [2]. It can also be used over long periods for nuclear power plant facilities and is very suitable for radiation shielding against gamma rays or neutrons. In the TRIGA PUSPATI reactor, they use high-density concrete as reactor shields, and the primary source of radiation used is neutrons.

However, after nearly thirty-nine years of service, the reactor is facing the effects of aging and is likely to have a reactor safety failure. History shows that concrete structures often deteriorate due to factors such as improper design, sloppy construction work, use of low-quality materials, exposure to high levels of chemicals that may endanger the safety of concrete structures and pose a risk to the safety and health of the public.

The problem involves neutron sources and gamma rays, which have a lower penetrating power than neutrons. The probability is that if irradiated with gamma rays, the concrete used can last for a long time. Even so, taking samples to see the condition of protective concrete against gamma rays in the TRIGA PUSPATI reactor is impossible. Therefore, research carried out with experimental methods to identify the type of concrete and aggregate mixture suitable for construction involves gamma ray sources.

The objective of this study is to determine the value of partial layers, tenths and linear attenuation coefficients from the rate of intensity of gamma rays that penetrate high-density concrete at different thicknesses and grades. In addition, this study was conducted to identify the optimal grade and mixture of concrete as a biological protection.

2. Study approach and results

This study uses ordinary high-density concrete that has a mixture of limestone aggregate and silica sand. The resulting four types of concrete grades are M15, M25, M35, M45 and what distinguishes these four concrete grades in terms of density and compressive strength. The samples were prepared at the Malaysian Nuclear Agency's Block 60 Laboratory. Cube samples (15x15x15 cm) were prepared and subjected to density and compression tests over 28 days. The four types of grades are then cut into five pieces in a thickness of 2.6 cm.

Concrete samples that have been prepared by mixing design as shown in the Table 1.

Table 1. Concrete samples that have been prepared by mixing design.

Grade	Water (kg)	Cement (kg)	Water/Cement (kg)	Gross Aggregate (kg)	Fine Aggregate (kg)	Compressive strength (N/mm2)
				Limestone	Silica sand	
M15	5.64	7.25	0.78	27.68	29.97	20.24
M25	5.64	9.74	0.58	28.96	26.20	32.89
M35	5.64	11.58	0.49	29.40	24.06	44.44
M45	5.64	13.37	0.42	29.25	22.28	39.93

Gamma ray counting was performed using a sodium iodide NaI (Tl) scintillation detector and a Single Channel Analyzer (SCA). The radioactive source used consists of radioactive elements Co-60 and Cs-137. After calculating the radiation intensity for each piece of concrete grade, the calculation of the linear attenuation coefficient, partial and tenth layer values is done with the following equation:

$$\begin{gathered} IX = I_0 e^{-\mu x} \\ HVL = \ln(2/\mu) \\ TVL = \ln(I_0/\mu) \end{gathered} \tag{1}$$

where *IX* is the number of photons after absorption, i_0 is the intensity before absorption occurs, μ is the linear attenuation coefficient and x is the thickness of the absorber of a material.

The result obtained from this study is that the radiation intensity decreases when the photon moves through the concrete. After doing the calculations, the linear attenuation coefficient increases as the concrete thickness increases for both sources (Co-60 and Cs-137). Previous studies also stated that the linear attenuation coefficient increases with increasing atomic number and increasing physical density of the absorbent material.

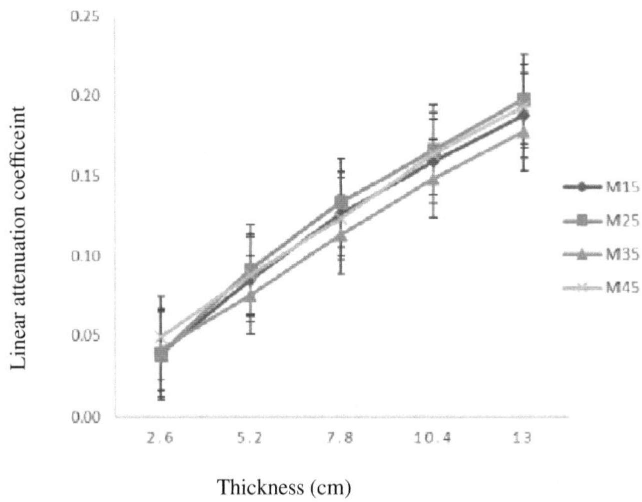

Figure 1. Relationship between thickness and linear attenuation coefficient of different concrete grades for the Cs-137 source

In characterizing the penetration of γ-rays in materials, the μ value plays an important role depending on the photon energy and the density of the shielding material [2] also proves linear attenuation coefficient obtain increases as the thickness increases but researchers have used magnetite and barite as a mixture of aggregates.

Figures 2 and 3 are graphs of the relationship between the thickness of concrete for the four grades and the difference in partial layer (HVL) and tenth layer (TVL) values for Co-60 and Cs-137 sources. If seen for both figures, the values of partial and tenth layers decrease while the concrete thickness increases. The standard deviation of the Co-60 source is greater than that of the Cs-137 source.

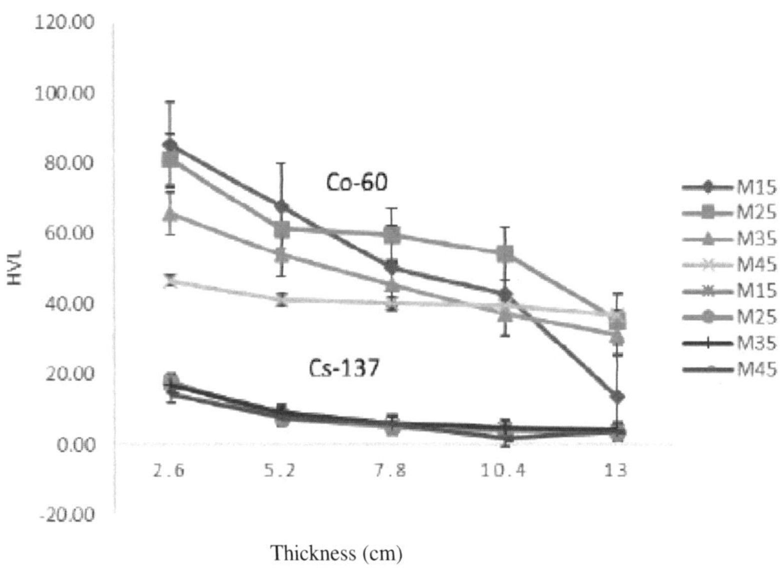

Figure2. Relationship between thickness and partial layer value difference (HVL) fourth concrete grade for Co-60 and Cs-137 sources.

Good absorption of radiation can be measured from values partial and tenth layers lesser. The relationship between the linear attenuation coefficient and this value is inversely proportional where when the attenuation coefficient increases, then the HVL and TVL values will decrease. It can be seen that there is a significant change between the HVL and TVL values obtained. It is possible that errors occurred when conducting the experiment and caused the standard deviation of some pieces of concrete grade.

The result of the graph shows that M35 grade is the optimal concrete grade as a gamma ray protector.M35 has higher durability and compression characteristics compared to other grades which is 44.44 N/mm2. This causes the concrete to absorb a large amount of radiation. This study has also made a comparison with previous studies. Among them is the study conducted by Sikora *et al.* [1]. A comparison of the current study with previous studies has been done and the results found that the value of the attenuation coefficient for M35, the source of Cs-137 is lower (0.159 cm-1) compared to this study (0.177 cm-1). The higher the value of μ, the better a material is as a radiation absorber. In other words, a higher μ value indicates a higher photon interaction or a lower penetration depth [8].

3. Conclusion

In conclusion, it was found that grade M35 is a better choice for use as a shielding material in nuclear facilities than other ordinary concretes because practicality and cheap.

4. References

[1] Sikora P, Abd Elrahman M, Horszczaruk E, Brzozowski P, Stephan D 2019 Incorporation of magnetite powder as a cement additive for improving thermal resistance and gamma-ray shielding properties of cement-based composites. *Constr. and Building Mater.* **204** 113–121.
[2] Ouda A S 2015 Development of high-performance heavy density concrete using different aggregates for gamma-ray shielding *Progress in Nuclear Energy* **79** 48-55.
[3] Baster I 1997 Calculation of radiation attenuation coefficients for shielding concretes. *Ann. of Nucl. Ener.* **24**(17) 1389-1401.
[4] Jignesh K, Vaniya S R 2015 Effect of use of silica sand as fine material in concrete *Micron.* **332**(694) 40-70.
[5] Jóźwiak-Niedźwiedzka D, Lessing P A 2019 High-density and radiation shielding concrete in developments in the formulation and reinforcement of concrete 193-228 (Woodhead Publishing).
[6] Mesbahi A, Ghiasi H 2018 Shielding properties of the ordinary concrete loaded with micro-and nano-particles against neutrons and gamma radiations *Appl. Radiat. and Isotopes* **136**, 27-31.
[7] Akkurt I, Basyigit C, Kilincarslan S, Mavi B, Akkurt A 2006 Radiation shielding of concretes containing different aggregates *Cem. Concr. Compost.* **28**(2) 153–157.
[8] El-Khayatt A M 2010 Radiation shielding of concretes containing different lime/silica ratios *Ann. of Nucl. Ener.*, **37**(7) 991-995.

Acknowledgement
This study was conducted partly using a research grant FRGS/1/2022/TK09/UKM/02/41/UKM/02/41

Radiation dosimetry for quality control of silicon wafer using electron beam

Shalina Sheik Muhamad[1,a)], Siti Zulaiha Hairaldin[1], Muhd Izham Ahmad[1], Shahrina Akma Mansur[1], and Noor Hasni M. Ali[1]

[1]ALURTRON, Malaysian Nuclear Agency,
43000 Bangi, Kajang Selangor

[a]shalina@nm.gov.my

Abstract. Dosimetry studies using electron beams were performed with energies of up to 3 MeV, 2mA with various trolley speeds on the silicon wafer. The silicon wafer came with diameters of 152.4 and 203.2 mm with a thickness of 0.5 mm. The irradiations were performed by utilising the EPS 3000 Cockcroft-Walton type 3.0 MeV, 90 kwatts electron beam (EB) accelerator. Plastic film dosimeter, cellulose triacetate (CTA) was used as a dosimeter with a UV-vis spectrophotometer as the readout system. The dose setting in this investigation was ~100 kGy for 152.4 mm diameter and ~160 kGy for 203.2 mm diameter, which was based on the requirements of the customer. The acceptable dose value was set (± 10%) for better performance of the silicon wafer. It was important to comply with the international standards protocol for dosimeters in radiation processing dosimetry. The results showed that the dose and temperature decreased with increasing the speed. The optimum dose for both diameter of wafers when irradiate at electron beam current of 2 mA and the voltage of 3 MeV is at speed of 0.89 m/min. Based on these studies, it was found that it is important to control the speed of the trolley as it affects the irradiation conditions (e.g., temperature and dose measurements).

1. Introduction

The use of radiation technology is widely used in industrial, medical and agricultural to modify materials. These include industries such as insulated wire as well as cable production, coatings, surface curing, sterilization of medical products, electronics as well as related fields (optoelectronics, photonics, sensors, etc.), disinfection, food preservation, wastewaters and flue gases treatment [1][2].

Various radiation sources are used to modify materials, such as high-intensity radioisotope sources, electron accelerators of various energies, and X-rays. During the treatment process using EB, the accelerating voltage and the beam current are important parameters. With an increase in the beam power, the kinetic energy of the electrons also increases, causing an increase in the beam penetration depth. The line speed is also important as it is directly interrelated to the irradiation time. Due to this movement, the heat is transferred to the volume of the material. An increase in the speed of the specimen will decrease the heat input [3-5].

EB irradiation of silicon wafer creates crystal defects within the material which act as deep traps. These deep traps capture charge carriers and in turn increase the resistivity/blocking voltage of silicon carbide [6]. However, there is a limitation of EB penetration in large diameter silicon wafer (>152.4 mm) restricting its application. Therefore, the feasibility of processing silicon wafer of large diameters with EB was explored. The relationship between electron beam parameters (energy), silicon wafer geometry (diameter), speed of trolley, temperature and dose distribution were investigated. This was because different irradiation conditions (energy, dose rate, etc.) on silicon wafers may cause different results.

2. Methodology

The experiments were carried out at the ALURTRON facility at the Malaysian Nuclear Agency. The facility provides irradiation services for local needs (commissioned in 1992) with energy of up to 3 MeV. Electron beam irradiation is the most efficient curing method for the wire and cable industry. Some wire and cable manufacturers had incorporated electron irradiation to provide high-quality heat-shrinkable tubes for the automotive industry. Local semiconductor companies also utilize the irradiation process to achieve the doping quality of the silicon wafer. ALURTRON also offers irradiation services to research and development institutes and universities.

The nominal electron energy used was 3 MeV. The beam current was kept constant at 2 mA during the experiments. The silicon wafer was positioned on the pallet of a trolley with the greatest care. The type of trolley used was a wire mesh type. The clearance between the accelerator output windows and the trolley was ~200 mm. The products were passed under a scanning horn several times to receive the desired dose. Irradiation was carried out in air at room temperature. The products were irradiated for 20 kGy in each cycle of irradiation thus the temperature did not increase significantly, which was to about 70 °C. The dose of 152.4 mm diameter in the range of 90–110 kGy and 203.2 mm diameter in the range of 144–176 kGy was based on the client's requirements. Details of the process parameters have been listed in Table 1.

Table 1. Process parameters

Factor	Value
Voltage	3 MeV
Current	2 mA
Trolley Speed	0.89, 0.92 and 0.94 m/min
Diameter of silicon wafer	152.4 and 203.2 mm

The chosen values of the trolley speed will influence the duration as well as the temperature of the electron beam irradiations and these in turn will affect the dose values. The product temperature was measured using the surface temperature indicating strips at each of the trolley passes. This process was based on the work instructions that were developed for quality systems [7]. (Figure 1 shows a procedure for the irradiation silicon wafer.

Cellulose triacetate (CTA) film dosimeter (FTR = 125, Fuji, Japan) was utilized to measure the absorbed dose of EB in the bunker. To ensure the film instability phenomena post-irradiation did not occur, the CTA was analysed after two hours of irradiation [8][9]. Figure 2 shows the placement of CTA film on the silicon wafer. After the irradiation, the CTA film was kept in the desiccator cabinet before the analysis. The dosimeter was read using a UV-VIS spectrophotometer (Genesys 5, France) at a 280 nm wavelength [10].

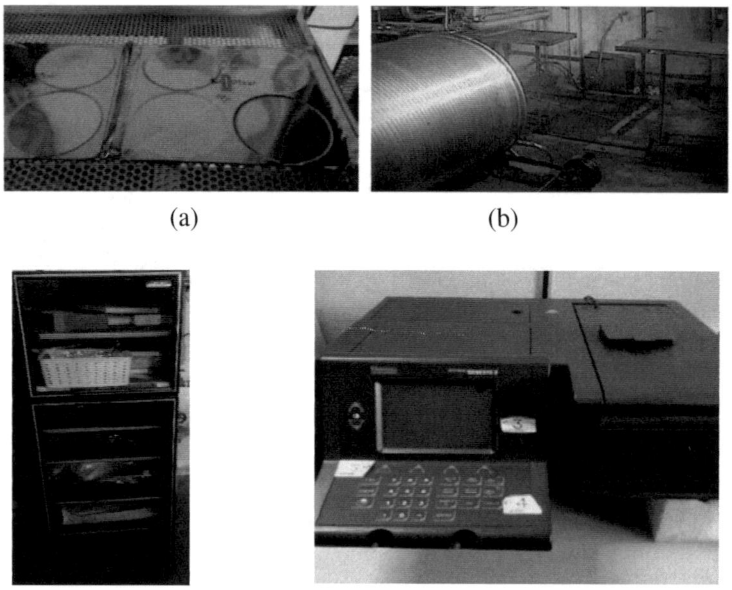

(a) (b)

(c) (d)

Figure 1. The procedure of the irradiation (a) The product preparation on the wire mesh trolley (b) irradiation of the products in the bunker (c) CTA film was kept in the desiccator cabinet for two hours after irradiation (d) analysis of the CTA film using UV-vis Spectronic Genesys 5.

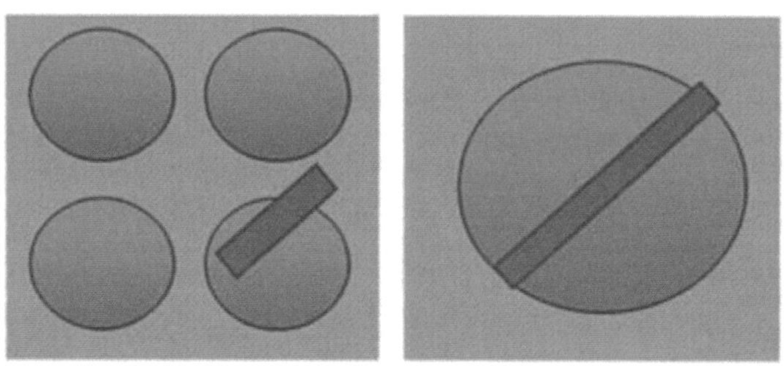

(a) (b)

Figure 2. Placement of CTA film (a) 152.4 mm wafer diameter and (b) 203.2 mm wafer diameter

3. Results and discussion

The effect of speed on dose and surface temperature has been shown in Table 2. In optimizing the dose in this study, it was of interest to find the best speed which fulfilled the optimum value for the dose. Based on Table 2, the suggested speed for both diameters of wafer was 0.89 m/min. To confirm the selection, we conducted a test on the 152.4 mm wafer diameter and the result was as shown in Table 3. The interaction effect of the speed with temperature and dose was shown in the graph in Figure 3. The graph shows that the dose and temperature decreased with an increase in speed. These results were in agreement with the findings of Valkov et al., (4) that the temperature decreases for a few degrees with an increase in the speed.

Table 2. The effect of process parameters on dose and surface temperature.

Diameter of wafer (mm)	Speed (m/min)	Min dose (kGy)	Max dose (kGy)	Mean Dose (kGy)	Tmax(°C)
	0.89	103.5	109.0	105.0	65
	0.92	96.9	102.5	99.3	65
152.4	0.94	93.5	97.7	95.2	54
203.2	0.89	158.0	170.5	165.0	65
	0.94	149.5	158.0	154.0	60

Table 3. The optimum dose for 152.4 mm wafer diameter.

Speed (m/min)	Min dose (kGy)	Max dose (kGy)	Mean Dose (kGy)	Tmax(°C)
0.89	95.9	102.0	98.7	60

The bulk energy density, which denotes how much energy is deposited per unit volume is defined as:

$$E_0 = \frac{UI}{vdh} \quad (1)$$

where E_0 is the bulk energy density [J/mm3], U is accelerating voltage [kV], I is beam current [mA], v is scanning speed [mm/s], d is the diameter of the electron beam (mm), and h is thickness of the sample [mm] [11]. The equation shows that at the constant electron beam power, the energy density decreases with an increasing speed. Increasing speed will reduce the bulk energy density, thereby reducing the dose on the top layers of the irradiated product. However, there were some limitations, such as that the equation (1) did not consider the changes in the thermophysical characteristics (thermal diffusivity, thermal conductivity, specific heat, density of the material) of the treated materials.

Based on the previous study by Mansur *et al.*, [12], the speed of the trolley was set at 0.94 m/min and the result was within the acceptable range. After the replacement of the ion pump of the EB Accelerator Tube, the speed of the trolley was set lower which was 0.89 m/min to ensure that the dose was within the range. This illustrated that when the trolley speed was reduced, the operational time was increased. Controlling trolley speed is a critical factor as it is a main factor observed by the quality and maintenance teams [13]. In the future, a series of tests with several trolley speeds should be performed, so that the calculation of the error can be carried out, followed by adjustment to the controller software of the servo motors. Scheduled maintenance, such as lubrication of the trolley system, could then also be carried out.

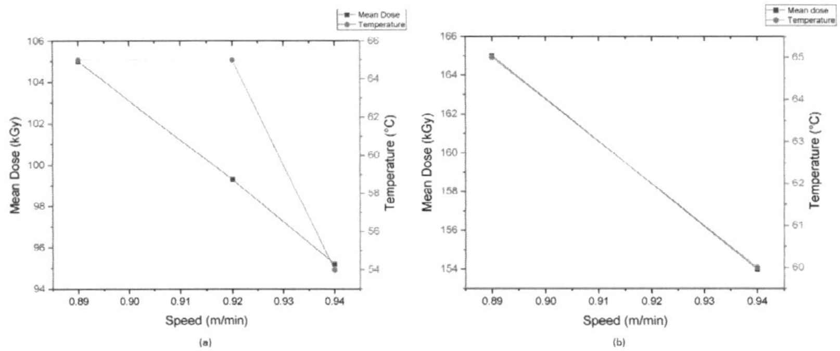

(a) (b)

Figure 3. The mean dose and temperature for the experiments for (a) 152.4 mm wafer diameter and (b) 203.2 mm wafer diameter

4. Conclusions

The dose and temperature on the silicon wafer surface subjected to continuous electron beam irradiation were investigated. This study revealed that the scanning speed significantly influenced the dose and temperature of the silicon wafer. The uniformity of the applied energy of an accelerator beam (exposure time) to a product would define the dose received. Thus, the study showed that it is important to have accurate knowledge of the dose distribution on the surface of the material. The other variables need to be studied comprehensively in the future. It is also important for routine maintenance of the trolleys because it leads to minimization of the energy cost and an increase in system performance.

5. References

[1] IAEA 2005 Emerging applications of radiation in nanotechnology *IAEA-Tecdoc-1438*.

[2] Sabharwal S. Electron Beam Irradiation Applications 2013 *Proc. of PAC2013* 745–8.

[3] Abou Elmaaty T, Okubayashi S, Elsisi H, Abouelenin S 2022 Electron beam irradiation treatment of textiles materials: a review *J. Polym. Res.* [Internet] **29(**4) Available from: https://doi.org/10.1007/s10965-022-02952-4

[4] Valkov S, Ormanova M, Petrov P 2019 Electron-Beam Surface Treatment of Metals and Alloys : Techniques and Trends Metals (Basel) **10(**1219) p 1–20.

[5] Valkov S, Ormanova M, Petrov P 2018 Surface Manufacturing of Materials by High Energy Fluxes *Adv. Surf. Eng. Res* [Internet] 69–87 Available from: http://dx.doi.org/10.1039/C7RA00172J%0Ahttps://www.intechopen.com/boos/advanced-biometric-technologies/liveness-detection-in-biometrics%0Ahttp://dx.doi.org/10.1016/j.colsurfa.2011.12.014

[6] Bullick A, Mauch D, Sullivan W, Dickens J 2014 Selective electron beam irradiation of high purity semi-insulating 4H silicon carbide substrates to characterize the effects on photoconductive semiconductor switch operation. *Dig. Tech. Pap. Int. Pulsed Power Conf. 2013* 4–8.

[7] ALURTRON 2019 Irradiation of silicon wafer product (EBM/WI/QA/06).

[8] JAERI1992 *CTA Dosimeter Manual: Radiation Film* FTR-125..

[9] Oproiu C, Toader D, Nemtanu M R, Popa R 2009 The use of the cellulose triacetate to determine the dose distribution of an electron beam. *Rom Reports Phys.* **54(**9–10) 861–7.

[10] ALURTRON 2018 Measurement of dose using UV-VIS spectronic (Genesys 5) (EBM/WI/QA/08).

[11] Li W, Ma R, Chen D, Yao Z, Song K, Yu L *et al.* 2021 The effect of continuous electron beam scanning process on the microstructure and geometry of U-5.5 wt%Nb alloy. *Nucl Instrum.*

Meth. Phys Res. Sect B Beam Interact with Mater Atoms [Internet] **496** 16–28. Available from: https://doi.org/10.1016/j.nimb.2021.03.019

[12] Mansur S A, Hairaldin S Z, Baharin R, Jusoh M S 2021 Kesan sinaran alur elektron ke atas CTA dengan penggunaan jenis alas troli berbeza ketika penyinaran silikon wafer *NTC 2021* 99.

[13] Fortes M Z, Schellember R V, Henry B 2015 Practical aspects regarding speed verification of servo motors driven conveyors *Recent Adv Electr Eng Electron Devices*.

Fire probabilistic safety assessment of control room in TRIGA PUSPATI research reactor

Zaredah Hashim[1], Nik Qurrotu Ain Ahmad Nasruddin[2], Mohd Huzair Hussain[1] Faizal Kunchi Mohamed[2], Phongsakorn Prak Tom[1], Julia Abdul Karim[1] and Mazleha Maskin[1,a)]

[1]Reactor Technology Center, Technical Support Division, Malaysian Nuclear Agency, MOSTI, 43000 Bangi, Kajang, Selangor, Malaysia

[2]Faculty of Science and Technology, Universiti Kebangsaan Malaysia, 43600, Bangi, Selangor, Malaysia

[a]mazleha@nuclearmalaysia.gov.my

Abstract. An internal Fire Probabilistic Safety Assessment (FPSA) study was conducted in the control room of the TRIGA PUSPATI (RTP) research reactor. This study aims to determine the likelihood of a fire occurring in the control room of RTP and the probability of fire occurring after an accident loss of instrumentation and control (LOIC) to the core damage frequency (CDF) value. The quantification results of the CDF are compared between LOIC with fire-induced (LOIC-F) and without fire-induced (LOIC). A methodical strategy to identify potential fire hazards that includes the precise identification of the design area, the size of any opening compartments, the location of electrical cabinets and fire suppression systems, as well as the placement of any safety-related equipment was applied. Included is Failure Mode and Effect Analysis (FMEA), which determines failure modes and designates repercussions for each scenario. This research will contribute on providing essential information on status of fire in RTP's control room as an insight of identifying any risk of fire scenario that may occur including investigation of current fire protection system sufficiency. In addition, a core damage end state of consequences of fire after an LOIC is also presented. The result of this research can be beneficial and creating awareness to the management as well as reactor operators in case of fire in the control room of RTP.

1. Introduction

PUSPATI TRIGA reactor (RTP) marked 40 years accident-free operation since its first criticality on June 28, 1982. Numerous activities were conducted to rejuvenate RTP to accelerate and increase operation and productivity especially on irradiation services as well as education and training. These activities include refurbishment [1], major modifications/upgrading: primary cooling system [2], instrumentation and control [3] and safety assessment [4][4][6] including human actions [7][8].

However, process of assessing the effectiveness of RTP in dealing with protection and mitigation of fire and smoke has not been studied yet. The fire safety and protection assessment of RTP in 2014 [9] merely qualitatively identifying the integrity of the system from the perspective of management and equipment, firefighting procedures and firefighting response teams. There is no quantitative evaluation was performed for instance the probabilistic safety assessment (PSA), which is widely known as an approach to estimate hazardous scenarios that may arise and to evaluate the seriousness of the accidents. Conversely, a study on probable causal of accidental fire ignition at the basement of RTP was conducted in 2020 [10]. Yet, this fire probabilistic safety assessment (FPSA) preliminary and pilot study to investigate the possibility of fire ignition as well as fire and smoke pace propagation have not been integrated with failure of any systems or components.

This paper is particularly observing the control room of RTP as another important location that assures main control defense of emergency incidents, consisting of various information, monitoring, and operation of the reactor itself. The control room of the RTP is a reasonable location for fire incidents, specifically electrical component failures [11][12]. Seeing this as a probe to enhance and optimize the current safety system of RTP, FPSA has been conducted to oversee any potential cause of fire. Besides, an updated PSA study level is necessary in favor of anticipating any changes when controlling RTP and studying the effectiveness of the current protection system [13].

Fire is one of the internal hazards that may happen in reactor. Several unpredictable cases such as Kashiwazaki Kariwa, a reactor in Japan that was on fire and broke out in an electrical transformer due to earthquake [14]. Another aftermath from fire incident led to a deterioration of electricity power generation was in Narora, North India [15]. Thus, this study is as an action of preserving the one and only reactor in Malaysia. Besides, Hassan *et al.* [16] executed a study on human actions (HAs) factor to scrutinize human reliability of operating RTP when fire happens. Therefore, a continuation of PSA study in control room as main supervision area is notably necessary.

The main objective of this study is to quantify the probability of fire occurrence and to assess the adequacy of current fire protection system at the control room. The scope of this study covers Level-I internal fire PSA, focusing only the reactor control room. The quantification results of the core damage frequency are compared between with and without fire-induced event.

2. Methodology

Methodology adapted in this study replicates as per work done in RTP basement [10].

2.1. Data collection and system identification

Identification and selection of components that may contribute to fire accidents are observed based on the provided Plant Walkdown sheet, supplied by the International Atomic Energy Agency (IAEA). This identification process gathered the information on which cables, openings and cable trunkings that are directly or indirectly hazardous to workers, public and plant itself. Component locations that comprise of fire extinguishers, openings (doors, windows, and ventilation system), or smoke detectors are attended as possible risk of fire. These locations were taken in a full layout of control room drawing. In obtaining the exact areas and coordinates of each component, measurements were carried out using BOSCH laser measuring device (Model No.: GLM 150C), iPad and measuring tape while at the same time existing schematic drawings are being referred. Finally, an accurate 2D and 3D drawings of the control room using Professional Autodesk Inventor 2023 (student's version) are drawn.

2.2. Data analysis

2.2.1. Identification of initiating event and failure mode and effect analysis (FMEA). In previous FMEA study, Hashim *et al.* [10], listed five (5) failure modes, namely: fire spark (FiS), no power (NP), fail to start (FS), operator action (OA) and no signal (NS). Meanwhile, four (4) consequences factors of the failure modes were identified, which are:

1) Leading to an initiating event (IE)
2) Affecting the ability of security functions to reduce IE
3) Influencing the operator's actions after the occurrence of IE
4) Leading to false functional movements that can cause harmful effects on people

IOP Conf. Series: Materials Science and Engineering 1285 (2023) 012015 doi:10.1088/1757-899X/1285/1/012015

2.2.2. Development of event and fault tree. An event tree with an internal fire event induced model is developed by modifying an initiating event (IE) loss of instrumentation and control (LOIC) that was developed in previous study [6] using event tree analysis (as depicted in Figure 1) for accident sequence analysis. Meanwhile, for system failure analysis, specifically on fire protection system, which consists of fire detector and fire suppression, are modelled using fault tree analysis technique. RiskSpectrum software is used in developing both event and fault trees analysis.

In quantifying the failure rate of fire by failure for both systems, several literatures were referred as described in Table 1. Meanwhile, for the quantification of fire IE, actual incident data is taken from the IAEA Integrated Reporting System for Research Reactor (IRSRR) [17] from year 1945 to February 2013. Due to the hassle in getting new password in accessing the database, data retrieved in previous study [18] were used with core damage end state remain the same (Table 2). No new data after February 2013 is involved.

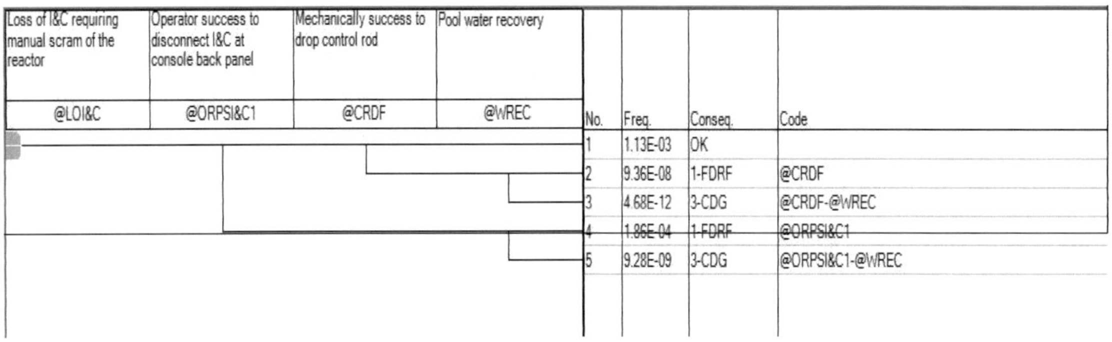

Figure 1. Loss of instrumentation and control (LOIC) [6].

Table 1. Fire detection and suppression failure rate.

Nos.	System	Failure rate	Source
1.	Fire suppression system: Pyrogen	20.0%	Vitali et al. [19]
2.	Fire detection system: Photoelectric	3.99%	Wisconsin Association of Home Inspectors [20]
		4.0%	Grosse et al. [21]

Table 2. Core damage end state.

ID	Description
1-FDRF	Reactor Protection System (RPS) failure, partial core uncovers, minor degradation during core refilling by external water
2-FDEA	RPS success, heat removal fails, complete core uncovers
3-CDG	RPS fails, rapid core uncovers, major degradation to the fuel in the core
4-RPWC	Release of pool water outside confinement
OK	No damage

3. Results and discussion

3.1. Data collection and system identification

Figure 2 depicted the RTP control room layout complete with cable trunkings, openings and electrical cabinets (RPS A, RPS B and DAC) as well as both fire detector and fire suppression systems. Most cables and trunkings were installed under floor of the control room to avoid trip hazards.

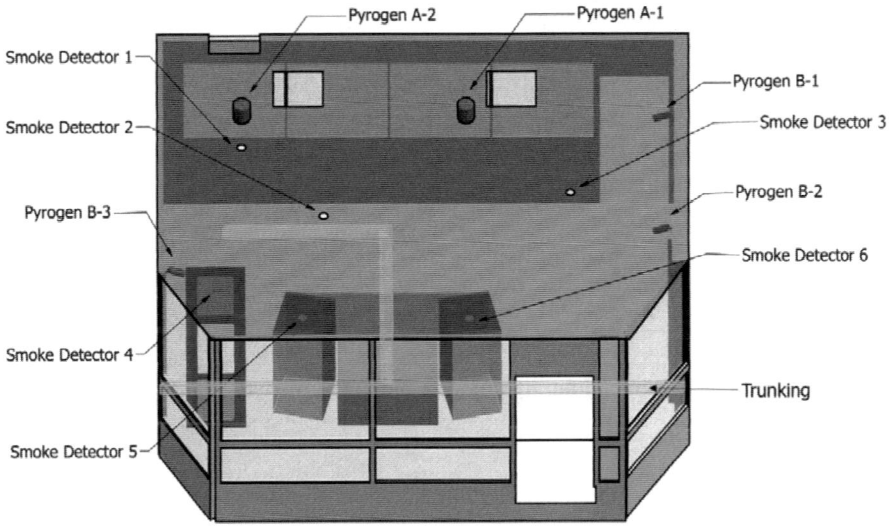

Figure 2. RTP control room layout.

3.2. Data analysis

Identification of Initiating Event and Failure Mode and Effect Analysis (FMEA)

Table 3 presents the lists of components with its failure mode and the consequences or effects to RTP if it fails.

Table 3. Results of FMEA.

Components	Failure Mode	Consequences or Effects	IE
1. Automatic Fire Detection: Smoke detector	FS	Operator missing compelling signal/ annunciator after the occurrence of fire	No
2. Manual Fire Detection: Detection by operator or reactor personnel	FS	Operator missing compelling signal/ annunciator after the occurrence of fire	No
3. Automatic fire suppression: pyrogen suppression device	FS	Operator missing compelling signal/ annunciator after the occurrence of fire	No
4. Manual fire suppression: intervention of personnel	FS	Operator missing compelling signal/ annunciator after the occurrence of fire	Yes
5. RPS A cabinet	FiS	Lead to an initiating event (IE)	Yes
6. RPS B cabinet	FiS	Lead to an initiating event (IE)	Yes
7. DAC C cabinet	FiS	Lead to an initiating event (IE)	Yes

*FiS Fire spark; FS: Fails to start

Based on the initial FMEA screening analysis, the 'Yes' value in Table 3, indicates two initiating events, which are fails to start (FS) and fire spark (FiS). However, these results will be re-evaluated based on the severity of the components in fire consequences over time.

3.2.1. Development of event and fault tree. The IAEA IRSRR recorded five (5) incidents on the internal fire which then were divided with the total exposure time of all research reactors in hour, resulted with a probability value of 8.0611E-08. This value is used as the IE of fire for this study.

Figure 3 depicted the altered IE for LOIC, with fire induced after the event operator successfully disconnected the instrumentation and control console back panel. LOIC-F showed four sequences of events with two end state of 1-FDRF and three end state of 3-CDG.

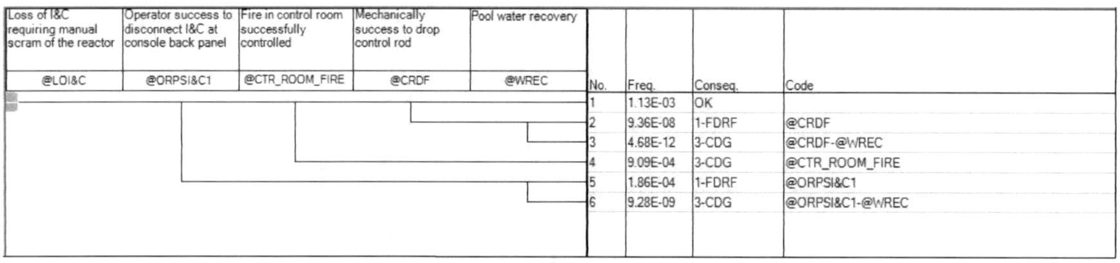

Figure 3. Event tree: Loss of instrumentation induced fire at sequence no. 2 (LOIC-F).

The fault tree analysis for Fire Protection System (FPS) is depicted in Figure 4, with the most conservative failure rates for both fire detection and suppression in Table 1 are used, 4.0% and 20.0% respectively. Control room FPS consists of two systems, which are Fire Detection System with three components and Fire Suppression System with five components. For this study, a conservative approach by assuming only one component fails, resulting with failure of FPS, hence OR gate is used.

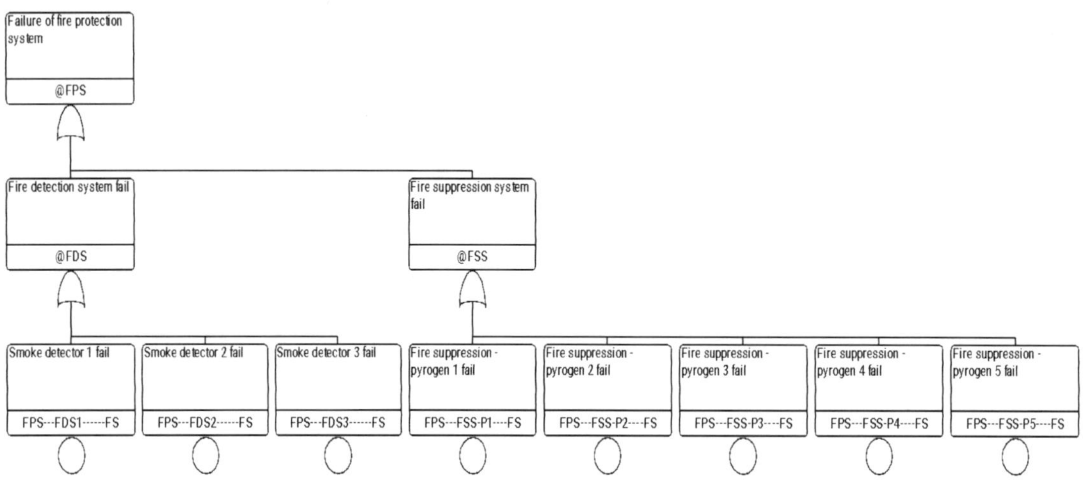

Figure 4. Fault tree: Fire Protection System (FPS).

From Table 4, the CDF value for LOIC and LOIC-F are 2.72E-08 and 9.40E-04 respectively.

Table 4. Core damage end state for both LOIC and LOIC-F

ID	LOIC	LOIC-F
1-FDRF	1.97E-04	1.97E-04
2-FDEA	1.01E-04	1.01E-04
3-CDG	2.72E-08	9.09E-04
4-RPWC	1.01E-04	1.01E-04
OK	8.99E-01	8.99E-01

4. Conclusion

It is concluded that, by inducing fire randomly as a sequence event in IE LOIC, the probability of core damage is 4 order magnitudes higher than the CDF from the original LOIC event tree. Therefore, the needs in making sure both fire detection and suppression works are mandatory. However, a further study by inducing fire in different sequence and other IE needs to be conducted to have an overview of internal fire impact to CDF at different IEs.

In assessing the adequacy of fire protection system, from reactor management point of view, FPS in the control room is adequate at present. However, as prevention is better than cure, or in this study, better than repair, therefore it is highly recommended that the RTP management should install several heat detectors in the control room as to strengthen the existing fire protection system. Supplementary PSA quantification in the addition of heat detectors will be the next follow-up project. In parallel, a continuation study of simulation in obtaining the speed and direction of fire and smoke propagation within and between compartments is also needed to be conducted to ensure the overall fire safety satisfactory at RTP.

5. References

[1] Masood Z, Zakaria M F and Hussin I A 2013 Refurbishment and modernisation of PUSPATI TRIGA reactor: Lessons learnt *15th Meeting of the Int. Group on Res. Reactors (IGORR15).* (Daejeon, South Korea).

[2] Lanyau T, Zakaria M F, Hashim Z, Abdul Farid M F and Kassim M S 2010 Replacement of PUSPATI TRIGA reactor primary cooling system and safety consideration J. of Nucl. and Related Tech. **7**(2) 112–114.

[3] Izhar A H and Mohd Idris T 2009 The development of a digital instrumentation and control system for PUSPATI TRIGA reactor console-requirement and experience *Nucl. Malaysia Tech. Conv. 2009 (*Malaysian Nuclear Agency).

[4] Ramli N, Karim J A, Md Razi H, Mustafa M K , MasenwaT N A, Sani S and Hussain M H 2020 Ageing assessment of biological shielding integrity for PUSPATI TRIGA Reactor IOP Conf. Ser.: Mater. Sci. and Eng. **785** 012051.

[5] Zakaria M F and Ramachandaramurthy V K 2017 Assessment and mitigation of power quality problems for PUSPATI TRIGA Reactor (RTP) AIP Conf. Proc. **1799** 020011.

[6] Brayon F C, Maskin M, Prak Tom P, Mohd Sarif A H S, Ramli Z, Zakaria F, Mohamed F, Aslam A, Lyubarskiy A, Kuzmina I, Hughes P and Ulses A 2014 Building competence for safety assessment of nuclear installations: Applying IAEA's safety guide for the development of a level 1 probabilistic safety assessment for the TRIGA research reactor in Malaysia *Probabilistic Safety Assess. and Manag. Conf. 12 (PSAM12)* (Sheraton Waikiki. Honolulu, Hawaii, USA).

[7] Phongsakorn P T., Ahmad Hassan S, Mazleha M, Nurul Husna Z A, Tonny L and Zaredah H 2016 Study on operator actions during the occurrences of undesirable events in PUSPATI TRIGA reactor *Res. and Dev. Sem. 2016* (Malaysian Nuclear Agency).

[8] Mohamed F, Mohd Sarif A H S, Yahaya R, Rahman I, Maskin M, Praktom P and Charlie F 2015 Operator reliability study for probabilistic safety analysis of an operating research reactor *Ann. of Nucl. Ener.* **80**, 409–415.

[9] Ab Rahim A N, Ligam A S,Ramli N, Zakaria M F, Hamzah N S, Prak P, Kassim M S and Masood Z 2014 Evaluation of fire safety and protection at PUSPATI TRIGA reactor (RTP) Res. and Dev. Sem. 2014 (Malaysian Nuclear Agency).

[10] Hashim Z, Kumar D, Zakaria M F, Hussain M H, Mohd Sarif A H S and Maskin M 2020 Internal fire probabilistic safety assessment preliminary work at research reactor TRIGA PUSPATI *IOP Conf. Ser.: Mater. Sci. and Engi.* **785** 012049.

[11] Jyung J M and Chang Y S 2021 Electrical fire simulation in control room of an AGN reactor *Nucl. Eng. and Tech.* **53**(2) 466–473.

[12] Lybarskiy A, Rozin V, Kouzmina I and Kazarians M 2004 Development of fire PSA methodology and its application in a study for Novovoronezh-5 NPP *Probabilistic Safety Assessment and Management PSAM7 — ESREL'04* ed. Spitzer C, Schmocker U, Dan V N (London: Springer-Verlag London Limited).

[13] IAEA 2001 *Applications of probabilistic safety assessment (PSA) for nuclear power plants (Tecdoc Series) No. 1200* (Vienna, Austria: International Atomic Energy Agency).

[14] Nagasawa K and Narabayashi T 2017 Seismic response and its analysis for components of Kashiwazaki-Kariwa Nuclear Power Plants in 2007 Niigata-ken Chuetsu - Oki Earthquake *Mech. Eng. J.* **4**(1) 16–00482.

[15] DivyaDeepak G 2012 Accident analysis of Narora fire accident *Int. J. of Sci. & Eng. Res.* **3**(2) 1–5.

[16] Mohd Sarif A H S, Maskin M, Prak Tom P, Brayon F C, Hlavac P and Mohamed F 2017 Operator response modeling and human error probability in TRIGA Mark II research reactor probabilistic safety assessment *Ann. of Nucl. Ener.* **102**, 179–189.

[17] IAEA 2014 Incident reporting system for research reactors (IRSRR) http://www-ns.iaea.org/tech-areas/research-reactor-safety/irsrr-home.asp [24 April 2014].

[18] Maskin M, Mohd Sarif A H S, Brayon F C, Phongsakorn P T, Zakaria M F, Ramli Z and Mohamed F 2018 Quantification of initiating event frequencies and component reliability data in level 1 probabilistic safety assessment at PUSPATI TRIGA research reactor *Ann. of Nucl. Ener.* **121** 22–28.

[19] Vitali J, Kibert C and Akers I 1996 Pyrogenic aerosol fire suppressants: Engineering of delivery systems and corrosion analysis *Proc. of the Halon Options Tech. Working Conf.* 7–9.

[20] Wisconsin Association of Home Inspectors 2015 Deadly differences: Ionization vs. photoelectric smoke alarms *WAHI Spring Conference, Ho-Chunk Casino and Conference Center, 13–14 March 2015.*

[21] Grosse L, DeJong J and Murphy J 1995 Risk analysis of residential fire detector performance *National Fire Project: Full Scale Research and Testing of Fire Detection Systems in a Residential Structure* (Texas A&M University).

Acknowledgement

This paper is produced under the Nuclear Malaysia internal budget fund No. NM-R&D-21-71. Additionally, as an extension of two other projects: (1) the Ministry of Science, Technology & Innovation (MOSTI) ScienceFund project (MOSTI 03-03-01-SF0179); and (2) IAEA – Norwegian Extra-Budgetary Program (NOKEBP) Fund, project name 'Applying PSA to Existing Facilities to Develop Transferable Skills in the use of PSA to evaluate nuclear power plant (NPP) Safety'. All projects are aiming on

IOP Conf. Series: Materials Science and Engineering 1285 (2023) 012015 doi:10.1088/1757-899X/1285/1/012015

developing a comprehensive technical work to master the PSA technology for a safe assessment of a nuclear facility through the practical application of the relevant recent IAEA Safety Standards on PSA to develop a PSA model for the only research reactor in Malaysia.

IOP Conf. Series: Materials Science and Engineering 1285 (2023) 012016 doi:10.1088/1757-899X/1285/1/012016

Assessment of elemental in plants acted as pollution biological indicator

Zalina Laili[1,a)], Lakam Mejus[1], Azian Hashim[1], Nurul Fairuz Diyana[1], Mohammad Syahiran Mustaffa[1], Muhammad Azfar Azman[1], Rahman Yaccup[1], Mohd Muzamil Mohd Hashim[1], Siti Aminah Omar[1], Shakirah Abd. Shukor[1] and Wilfred Paulus[1]

[1]Malaysian Nuclear Agency, Bangi, 43000 Kajang, Selangor, Malaysia

[a]liena@nm.gov.my

Abstract. Direct exposed to sources of heavy metal pollution in the environment have increased the ability of a certain plant to adopt and grown in this kind of environment. Heavy metals concentration in plants grown in the Pasir Gudang industrial area were determined using neutron activation analysis (NAA) and X-ray fluorescence (XRF). The focus of this paper is to evaluating the potential use of plants in industrial areas as biological indicators to air and soil pollution. Nine plants species were collected based on the plants available at factories in the Pasir Gudang, Johor industrial area. The concentration of Fe, Zn, Co, Cr, As, Mn, Sb, Pb, Ni, Cd, and Cu in all plants were in the range of 0.186 to 482 mg kg-1 (dry weight). The results revealed the significant higher concentration of Al, Fe, Zn dan Mn in all polluted plants. Higher concentrations of Al, Fe and Zn were recorded in leaves of Acacia mangium leaves compared to other plant species. Our findings suggest that Acacia mangium plant is suitable as a tracer plant for a certain heavy metal absorption. More comprehensive monitoring is needed to understand the accumulation of heavy metal in the plant collected in Pasir Gudang Industrial area.

1. Introduction

Plants are useful for assessing the state of the environment conditions and for monitoring environmental responses on any pollution sources. In recent years, there has been increased awareness of the role of plants in environmental pollution. Plants growing in industrial and urban areas can serve as bioindicators or biomonitors of pollution [1]. Various researchers have stated that the sources of heavy metal pollution in the environment are due to anthropogenic activities. In general, anthropogenic activities such as the steel and iron industries, mining, smelting, transportation and agricultural activities have led to increased exposure of the environment to heavy metals [2]. Many heavy metals such as Fe, Cu, Zn, Mn and Ni are considered crucial for plant growth [3][4]. However, elevated concentrations of heavy metals cause severe toxicity in plants, triggering physiological and biochemical changes [5]. The concentration of the metals in the leaf of woody and herbaceous plants growing near emission sources can reflect the level of pollution in these areas [2]. The tree leaves grown in urban area were change chemically and physiologically after exposure to various level of air pollution [6].

In Malaysia, the pollution of the Kim Kim River in Pasir Gudang has attracted public and media attention in 2019 after thousands of people were affected by the illegal dumping of chemical waste into the river. Pasir Gudang is well known as an industrial area located in Mukim Plentong, Johor Bahru District, Johor, Malaysia. Major industries established here are transport and logistics, shipbuilding, petrochemicals, food productions, container services and oil palm storage and distribution. Several studies in the literature that have detected higher metal concentrations in sediments and shellfish (Perna Veridis) in Kampung Pasir Putih near Pasir Gudang ([7], [8], [9], [10]). However, no studies on the concentration of heavy metals in a crop grown in the Pasir Gudang industrial area have been found in the literature so far. We can better understand the bioavailability of metals to biomonitors by using a set of biomonitors representing different levels in the intertidal ecosystem.10]. Thus, the purpose of this study is to provide an assessment of heavy metal accumulation in plants grown in the Pasir Gudang industrial area that may have the potential as a bioindicator of heavy metals pollution.

2. Material and methods

2.1. Sampling location

Sampling sites were located in and around industrial area in Pasir Gudang as shown in Fig. 1. The key industries at this area include electrical and electronics, food products, chemical, plastics and engineering-based industries.

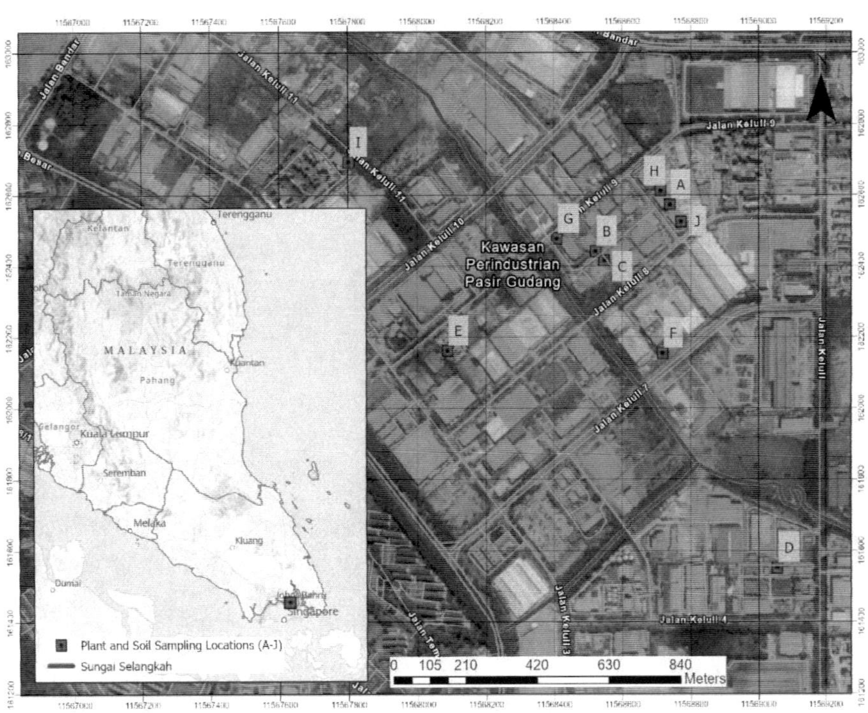

Figure 1. Sampling Location.

2.2. *Plant and Soil Sampling and Preparation*

Figure 2 showed the selected plant species collected for this study. The samples were collected based on the plants available at factories in the Pasir Gudang, Johor industrial area. The soil samples were also collected from 2 - 10 cm of the surface soil. The collected plant samples were oven-dried at 50ºC for 48 hours, crushed, homogenized, and sieved at 1 mm particle size. The soil samples were oven-dried at 50 ºC for about 10 -15 h to obtain a constant dry weight and then sieved to pass through a 30-micron mesh sieve.

Figure 2. Selected plants species collected at the Pasir Gudang Industrial Area.

2.3. *Analysis of heavy metals*

Analysis of Al, Fe, Zn, Co, Cr, As, Mn and Sb in soil and plants was carried out by neutron activation analysis (NAA). Standard reference materials (SRMs) (NIST SRM 1573a Tomato Leaves, NIST SRM 2709a San Joaquin Soil and IAEA-SL1 Lake Sediment) were irradiated along with the sample, standard and blank to ensure accuracy and precision of the data. All samples and standard reference materials were irradiated in the 750 kW PUSPATI TRIGA Mark II research reactor of the Malaysian Nuclear Agency with a thermal neutron flux of 4.0×1012 n cm-2 s-1. The irradiation and counting procedure were carried out according to [11]. A high-resolution coaxial Ortec hyperpure germanium (HPGe) detector with a resolution of 1.9 keV at 1332.5 keV of 60Co was used to count the samples. The Gamma Vision analysis software was used to analyse the spectral data of the samples.

For determination of Pb, Ni, Cd, and Cu in plants was carried out by X-ray fluorescence (XRF) Rigaku Supermini 200 (Tokyo, Japan) type benchtop sequential spectrometer equipped with an air-cooled 200 W Pd target X-ray tube (50 kV, 40 mA). Instrument performance was verified by scanning certified reference materials from the National Institute of Standards and Technology (NIST) (NIST 1515 Apple Leaves and NIST 2711a Montana II Soil). Before measurement, 2 g of powdered sample was weighed and put into the sample cell. The polypropylene film (6μm) was placed over the sample cell then covered with a pre-vented lid. The sample was exposed to a high-energy photon from an X-ray tube that induced the transition between atomic orbital and contributes to the emission of a photon. By measuring the energy and intensity (count rate) of these photons, qualitative and quantitative information is obtained.

2.4. *Quality control and quality assurance of measurement*

The Standard reference material namely Tomato Leaves 1573a, San Joaquin Soil 2709a and Lake Sediment IAEA-SL, Apple Leaves 1515 and Montana II Soil 2711a were used as quality control and quality assurance in the analytical method analysis. The calculation of relative bias (%) is describe in equation (1):

$$\text{Relative} - \text{bias} \frac{C_{analysed-C_{certified}}}{C_{Certified}} \text{ X } 100\% \tag{1}$$

where $C_{analysed}$ is the concentration of analaysed value, $C_{certified}$ is the concentration of certified value.

The accuracy of the analytical techniques was evaluated using a Z-score calculation [11]:

$$Z - score = \frac{x-c}{\sqrt{u_x{}^2+u_c{}^2}} \tag{2}$$

where x is the analytical result, c is the certified value, u_x is uncertainty of analytical result and u_c is uncertainty of certified value. The uncertainty of the analytical results is based on counting statistics and the uncertainty of certified value based on the certificate. For acceptance of the results: $-2 < z < 2$ is anticipated. However, is $z < -3$ or $z > 3$, is consider that the result is "out-of-control" and correction action will be taken.

2.5. *Heavy metals bioaccumulation*

Bioaccumulation Factor (BAF) is defined as a ratio of metal concentration in plant to concentration of metal in soil [12]:

$$BAF = \frac{C_{plant}}{C_{soil}}, \quad (3)$$

where C_{plant} is the concentration of metal in plant (mg kg-1) and C_{soil} is the concentration of metal in soil (mg kg-1). If the BAF value is less or equal to 1.00, it indicates that plant can only absorb but not accumulate the metal. However, if the BAF value is more than 1.00, it indicates that the plant may have potential to accumulate the metal [13].

2.6. *Statistical analysis*

Statistical analysis of the data collected was carried out using the Statistical Package for Social Science (SPSS version 22.0). The ANOVA test was used to determine the significant difference within and between groups, considering a significance level of less than 5% ($P < 0.05$). The relationships between the concentration of heavy metal in plant and soil were statistically tested using Pearson product-moment linear correlation coefficient. In this, we define the null hypothesis H_o as 'a linear relationship is non-existent (or uncorrelated) between the two parameters'. The calculated probability value is the probability of being wrong if we reject H_o.

3. Results and discussion

3.1. *Quality assessment*

The certified and measured value of all SRM, percentage of recovery, relative bias and Z-score are shown in Table 1. The analysed of SRM showed that the results are within the acceptance criteria. The recovery and relative bias of SRM ranged 82.6 to 114.1 % and -0.1 to 16 %, respectively.

3.2. *Heavy metals in the soil*

Table 2 shows the concentration of Al, Fe, Zn, Co, Cr, As, Mn, Sb, Pb, Cu, Ni, and Cd in the soil samples collected from the selected factories compound in the study area. The concentration of heavy metals measured in the soil samples ranging from 0.64 to 183,256 mg kg^{-1} (dry weight). The highest heavy metal concentration in the soil samples is Al (183,256 mg kg^{-1}) collected in the compound of a metal industry (Factory J). The typical range of naturally occurring Al concentration in Malaysian soils is between 33,500 and 53,900 mg kg^{-1}[15]. The results indicate that the Al concentration was three times the naturally occurring concentration reported by DOE. The Zn concentration (64.59 3 mg kg^{-1}) in the soil samples collected from Factory D (Prescribed Premises Scheduled Wastes) was slightly above the normal concentration range in Malaysian soils (6.9 to 54.3 mg kg^{-1}). The results also show that the concentrations of Mn (248 to 4213 mg kg^{-1}), Cr (19.1 to 468.78 3 mg kg^{-1}) and Pb (21.1 to 88.1 3 mg kg^{-1}) in the soil samples collected from the factories are above the normal range in Malaysian soils (3.95 to 3.99 mg kg^{-1} for Mn; 0.02 to 14.4 mg kg^{-1} for Cr; 0.18 to 36 mg kg^{-1} for Pb). The results indicate that the heavy metal concentrations in the soils of the factories were influenced by the activities in and around the area.

Table 1. Result of analysis of Standard Reference Material (in mg kg^{-1}).

Elements	Measurement value	Certified value	Recovery (%)	Relative Bias (%)	Z-Score
Analysed by NAA					
Tomato Leave 1573a					
Al	609 ± 0.49	598 ± 7.1	100.10	1.83	1.58
Ba	64 ± 16.7	63	100.60	1.58	0.06
Ca	50500 ± 5.2	50450 ± 550	100.1	0.10	0.09
Fe	318.24 ± 21.6	367.5 ± 4.3	86.77	-13.40	-2.23
K	27000 ± 5.8	26750 ± 480	99.89	-0.12	-0.03
La	2.04 ± 2.3	2.3	88.8	-11.22	-0.24
Sm	0.22 ± 0.03	0.19	114.1	14	0.94
San Joaquin Soil 2709a					
Ba	1002.40 ±221	979 ±28	102.39	2.39	0.11
Cs	5.85 ±0.87	5 ± 0.1	116.97	16.00	0.97
Co	14 ±1.4	12.8 ±0.2	109.29	9.29	0.86
Fe	22736.60	33600 ±0.07	82.60	-17.45	-8.78
Rb	±607.5	99.3 ±3.0	111.95	11.94	1.01
Sc	110.83 ±11.31	11.1 ±0.1	106.32	6.32	0.47
Zn	11.80 ±1.5	103 ±4	89.96	-10.04	-1.30
	92.66 ±6.9				
Analysed by XRF					
Apple Leaves 1515					
Al	291	284.5 ±5.8	102.28	2.28	1.10
Ba	52	48.8 ±2.3	108.33	8.33	1.59
Ca	15327	15250 ±100	100.50	0.50	0.77
Cu	6	5.69 ±0.13	105.4	5.40	0.31
Fe	82	82.7 ±2.6	99.15	-0.84	-0.25
Mn	55	54.1 ±1.1	101.66	1.66	0.61
Sr	25	25.1 ± 0.43	99.60	-0.40	-0.07
Montana II Soil 2711a					
As	108	107 ± 5	100.93	-0.93	0.20
Co	10	9.89 ± 0.18	101.11	1.11	0.11
Cs	7	6.7± 0.2	104.48	4.47	0.29
Mn	684	675 ±18	101.33	1.33	0.50
Ni	23	21.7 ± 0.7	105.99	5.99	1.07
Rb	122	120 ± 3	101.67	1.66	0.63
V	81	80.7 ± 5.7	100.37	0.37	0.051

Table 2(a). Heavy metal concentration in soil samples at the factories area and typical range of metal concentration in Malaysian soils.

Factory Type	**Concentration (mg kg^{-1})**							
	NAA							
	Al	Fe	Zn	Mn	As	Cr	Co	Sb
Factory A- Chemical	5856	19430	36.67	288	3.87	37.52	1.85	0.9
Factory B – Metal Manufacturing	43429	13337	45.95	248	3.32	80.13	2.31	0.88
Factory C - Steel Fabrication	34544	35945	31.47	325	4.68	468.78	3.95	2.14
Factory D - Prescribed Premises Scheduled Wastes	62582	21684	64.59	207	9.51	119	2.31	1.64
Factory E- Rubber	45925	18448	53.19	<0.01	2.01	201.94	2.76	0.64
Factory F- Foods Processing	54736	13316	48.98	229	5.9	19.06	<0.1	0.84
Factory J - Prescribed Premises Scheduled Wastes	59827	25667	53.86	421	8.11	61.12	3.69	1.28
Factory H - Metal Finishing and Electroplating	48837	14288	38.88	<0.01	6.13	50.96	<0.1	1.05
Factory I - Metal Manufacturing	183256	11262	30.06	<0.01	5.05	21.82	1.04	1.17
Factory J - Metal Finishing and Electroplating	167917	30170	36.04	<0.01	4.95	451.91	6.1	2.12
Typical range of natural occurring metals concentrations in Malaysian soils	33500 - 53900	301 - 44500	6.9 - 54.3	3.95 - 3.99	1.1 - 43.0	0.02 - 14.40	13.9 - 11.9	-

Table 2(b). Heavy metal concentration in soil samples at the factories area and typical range of metal concentration in Malaysian soils.

Factory Type	Concentration (mg kg^{-1}) XRF			
	Pb	Cu	Ni	Cd
Factory A- Chemical	30.2	73	20	30
Factory B – Metal Manufacturing	34.6	75	140	33
Factory C - Steel Fabrication	53.7	457	299	30
Factory D - Prescribed Premises Scheduled Wastes	28.2	62	22	33
Factory E- Rubber	23.2	67	23	30
Factory F- Foods Processing	21.1	62	19	34
Factory J - Prescribed Premises Scheduled Wastes	47.0	87	25	45
Factory H - Metal Finishing and Electroplating	20.2	74	47	40
Factory I - Metal Manufacturing	78.7	81	19	40
Factory J - Metal Finishing and Electroplating	88.1	250	130	39
Typical range of natural occurring metals concentrations in Malaysian soils	0.18 - 36.0	4.0 - 19.8	0.7 - 28.9	0.09 - 11.90

3.3. Heavy metals in plants

The concentrations of heavy metals in the selected plant leaves collected from the factory premises are shown in Table 3. The heavy metal concentrations measured in the plant leaves ranged from 0.186 to 482 mg kg^{-1} (dry weight). Statistically, Table 4 shows that there was a significant difference in the mean heavy metal concentration between plant and soil ($p = 0.01$). The statistical result of Pearson correlation (Table 5) shows a weak correlation (0.214) between heavy metal concentration in the plant and soil. This result suggests that there might be other factors that have a stronger influence on the heavy metal concentration in the plants.

Heavy metals are considered trace elements due to their very low concentrations (from the ppb range to < 10 mg kg^{-1}) in various environmental matrices [16]. However, our results show that the concentration of Al, Fe, Zn and Mn exceeded 10 mg kg^{-1} in most of the plant leaves sampled from the selected factories compound, especially at the metal and steel processing factories. This observation would suggest that anthropogenic activities may influence the concentration of heavy metals in plants. The urban and industrial activities can increase the heavy metals present in the atmosphere [17]. Airborne lifted heavy metals could be deposited on the surface of the soil and plants leaves. Plants also can absorb the heavy metal through their roots and vascular system and translocate them into vegetative tissues and other tissues, as well as penetrated through the stomata and accumulate in the leaf tissue [2]. Therefore, it can be assumed that the enhanced heavy metals concentration in leaves from the study area was filtered from the air. Moreover, there are several other factors can affect the concentration of heavy metals in plants leaves such as metal concentrations in soil, soil pH, plant species, solubility, the chemical form of elements, soil pH, organic matter content, texture and cation exchange capacity (CEC) of the soil ([13], [18]). Additionally, our results show that the concentration of Al and Fe in *Acacia mangium* was exceeding the normal toxic limit (> 300 mg kg^{-1}), which is considered as heavily toxic. The toxic level of metals in plants is 300 mg kg^{-1}[19].

Table 3 also shows that higher Al contents were found significantly in the leaves of *Acacia mangium* (Factory I-Metal Manufacturing), *Euphorbia tithymaloides* (Factory J - Metal Finishing and Electroplating*)*, *Baphia nitida* (Factory F- Foods Processing) and Ficus microcarpa L.f. (Factory B - Metal Manufacturing). The Al concentration measured in the leaf samples taken in these factories ranged from 70 to 482 mg kg^{-1} (dry weight). Moreover, the Al concentration in the soil around the plants was also high, ranging from 43.429 to 183.256 mg kg^{-1} (dry weight). Thus, our results suggest that Al concentrations in plant leaves and soils may be related to contamination from metal processing in factories.

Fe concentration in the plant samples ranging from 70 to 434.5 mg kg^{-1} (dry weight), and significantly exceeded the toxic limit (> 300 mg kg^{-1}) as well as above the permissible limit set by WHO (20 mg kg^{-1}) [20], especially in *Acacia mangium* and *Euphorbia tithymaloides* grown in the metal processing factories. Although Fe is an essential nutrient for various physiological processes in plants [2], over-absorb by plant can damaged the plant tissues. Other heavy metals measured in the results show that the range of Zn concentration in plant leaves is between 1.0 and 226.5 mg kg^{-1} (dry weight). The highest Zn levels were found in *Acacia mangium* (226.5 mg kg^{-1} dry weight). Although Zn is an essential nutrient for plants, the requirement of Zn in plants is around 15-20 mg kg^{-1} [21]. The range of Mn concentration in the plant leaves is between 2.4 and 158 mg kg^{-1}. Mn was found to be the highest in the *Baphia nitida* (Factory F- Foods Processing) and the lowest in the leaf sample of *Ficus microcarpa L.f* (Factory B – Metal Manufacturing). Mn is one of the 17 essential elements for plant growth and reproduction and it is required in small quantities by plants ([21], [22]).

Table 3(a). Heavy metal concentration in leaves of plant grown in Pasir Gudang industrial area.

Plant Species	Factory Type	**Concentration (mg kg^{-1})**							
		NAA							
		Al	Fe	Zn	Mn	As	Cr	Co	Sb
Pandanus pygmaeus	Factory A Chemical	184	102.5	26.5	18.0	0.615	1.715	<0.1	0.186
Ficus microcarpa L.f.	Factory B – Metal Manufacturing	222	265.5	1.0	2.4	0.720	1.850	<0.1	0.545
Osmoxylon Lineare	Factory C - Steel Fabrication	110	268.0	55.5	16.4	0.148	4.185	0.136	0.480
Tabebuia rosea	Factory D - Prescribed Premises Scheduled Wastes	147	70.0	23.0	14.2	0.429	0.950	0.198	0.505
Acacia mangium	Factory E- Rubber	103	143.5	32.5	55.7	<0.1	1.275	0.163	0.570
Baphia nitida	Factory F- Foods Processing	249	138.5	42.5	158.3	0.505	2.315	<0.1	0.490
Ixora sp	Factory G - Prescribed Premises Scheduled Wastes	70	88.5	80.5	19.0	<0.1	1.470	<0.1	0.540
Murraya paniculata	Factory H - Metal Finishing and Electroplating	105	254.5	32.5	54.5	0.154	2.445	0.225	0.565
Acacia mangium	Factory I - Metal Manufacturing	482	434.5	226.5	94.3	0.261	3.095	0.261	0.313
Euphorbia tithymaloides	Factory J - Metal Finishing and Electroplating	275	357.5	183.5	28.9	0.780	5.275	0.718	0.558

IOP Conf. Series: Materials Science and Engineering 1285 (2023) 012016 doi:10.1088/1757-899X/1285/1/012016

Table 3(a). Heavy metal concentration in leaves of plant grown in Pasir Gudang industrial area

Plant Species	Factory Type	Concentration (mg kg^{-1})			
			XRF		
		Pb	Cu	Ni	Cd
Pandanus pygmaeus	Factory A Chemical	<1	5	1	<1
Ficus microcarpa L.f.	Factory B – Metal Manufacturing	<1	6	1	<1
Osmoxylon Lineare	Factory C - Steel Fabrication	<1	6	2	<1
Tabebuia rosea	Factory D - Prescribed Premises Scheduled Wastes	<1	7	1	<1
Acacia mangium	Factory E- Rubber	<1	7	1	<1
Baphia nitida	Factory F- Foods Processing	<1	7	1	<1
Ixora sp	Factory G - Prescribed Premises Scheduled Wastes	<1	9	1	<1
Murraya paniculata	Factory H - Metal Finishing and Electroplating	<1	6	1	<1
Acacia mangium	Factory I - Metal Manufacturing	<1	7	1	<1
Euphorbia tithymaloides	Factory J - Metal Finishing and Electroplating	<1	5	1	<1

Table 4. ANOVA analysis of heavy metal concentration between plant and soil

	Sum of Squares	df	Mean Square	F	Sig.
Between Groups	3852736864.220	1	3852736864.220	11.476	.001
Within Groups	79904308608.621	238	335732389.112		
Total	83757045472.841	239			

Table 5. Pearson Correlation analysis heavy metal concentration between plant and soil.

		Sample	Value
Sample	Pearson Correlation	1	.214**
	Sig. (2-tailed)		.001
	N	240	240
Value	Pearson Correlation	.214**	1
	Sig. (2-tailed)	.001	
	N	240	240

3.4. Bioaccumulation Factor (BAF)

Table 6 shows the bioconcentration factor (BAF) for heavy metals calculated from the results obtained in plant species collected from the factories. Low BAF (BAF < 1) of Al, Fe, Mn, As, Cr, Co, Sb, Cu and Ni were found in most of the different plant species, suggesting that the plants can only take up the metals but not accumulate them. Despite the higher concentration of Al and Fe in the soil, the BAF for Al and Fe in plants was found to be low. This could be related to the plant's strategy for detoxification mechanisms [24]. However, the BAF value for Zn was above 1 in *Osmoxylon Lineare*, *Acacia mangium*, *Ixora* sp and *Euphorbia tithymaloides*, suggesting that these plant species accumulate Zn. The highest BAF of Zn was found in *Acacia mangium* (7.53). In general, the bioavailability and the accumulation of heavy metals in plant depends on several factors such as the type of soil and properties, metal concentration and properties, plant species and urbanization level. Thus, comprehensive studies should be conducted at Pasir Gudang industrial area to understand the accumulation of heavy metals in plant.

Table 6. Heavy metals bioconcentration factor (BAF) of different plants species at Pasir Gudang industrial area.

Plant Species	Al	Fe	Zn	Mn	As	Cr	Co	Sb	Cu	Ni
Pandanus pygmaeus	0.0033	0.0053	0.723	0.063	0.16	0.046	-	0.21	0.068	0.05
Ficus microcarpa L.f.	0.0051	0.012	0.022	0.010	0.22	0.023	-	0.62	0.08	0.0071
Osmoxylon Lineare	0.0032	0.0075	1.76	0.050	0.032	0.0089	0.034	0.22	0.013	0.0033
Tabebuia rosea	0.0023	0.0032	0.36	0.069	0.045	0.0078	0.086	0.31	0.12	0.045
Acacia mangium	0.0022	0.0078	0.61	-	-	0.0063	0.059	0.89	0.10	0.043
Baphia nitida	0.0045	0.010	0.87	0.69	0.086	0.12	-	0.58	0.11	0.052
Ixora sp	0.0012	0.0034	1.49	0.045	-	0.024	-	0.42	0.10	0.04
Murraya paniculata	0.0022	0.012	0.84	-	0.025	0.048	-	0.54	0.081	0.021
Acacia mangium	0.0026	0.039	7.53	-	0.052	0.14	0.25	0.27	0.086	0.052
Euphorbia tithymaloides	0.0016	0.012	5.09	-	0.16	0.012	0.12	0.26	0.02	0.0077

4. Conclusion

Heavy metal concentrations were determined in the soil and plants of selected factories in the Pasir Gudang industrial area. The significantly higher concentrations of Al, Fe, Zn and Mn in all the polluted plants indicate that the industrial activities affect the heavy metal concentration in the environment. In this study, *Acacia mangium* was found to have the highest heavy metal concentration and can be used as a biological indicator of heavy metal pollution in an industrial area. However, to better understand the accumulation of heavy metals in plants grown in an industrial area, more comprehensive monitoring is needed.

5. References

[1] Simon E, Braun M, Vidic A, Bogyo D, Fabian I and Tothmeresz B 2011 Air pollution assessment based on elemental contentration of leaves tissue and foliage dust along an urbanization gradient in Vienna *Environ. Pollut.* **159** 1229-1233.

[2] Salih Z and Farhad A 2019 Heavy metals accumulation in leaves of five plant species as a bioindicator od steel factory pollution and their effects on pigment content. *Pol.J.Environ. Stud.* **28**(6) 4351-4358.

[3] Dogan Y, Unver M C, Ugulu I, Calis M and Durkan N 2014 Heavy metals accumulation in the bark and leaves of *Juglans regia* planted in Artvin City. *Biotechnol. & Biotechnol. Equip.* **28**(4) 643-649.

[4] Juson A E D S, Martinez M K M and Ching J A 2016 Accumulation and distribution of heavy metals in Leucaena leucocephala Lam and Bougainvillea spectabilis wild plant systems *J. Exp. Biol. Agric.* Sci. **4** 1-6.

[5] Singh S, Parihar P, Singh R, Singh V P and Prasad S M 2016 Heavy metal tolerance in plants: role of transcriptomics, proteomics, metabolomics, and ionomics. *Frontiers in Plant Sci.*. **6**(1143) 1-36.

[6] Zadeh A R K, Veroustraete F, Buytaert, J A, Dirckx J, Samson, R 2013. Assessing urban habitat quality using spectral characteristics of Tilia leaves. *Environ. Pollut.* **178** 7-14.

[7] Abdullah A R, Woon W C and Bakar R A 1996 Distribution of oil and grease and petroleum hydrocarbons in the Straits of Johor, Peninsular Malaysia. *Bull. Environ. Contam. Toxicol.* **57**(1) 155–162.

[8] Zulkifli S Z, Ismail A, Mohamat-Yusuff F, Arai T, and Miyazaki N 2010. Johor Strait as a hotspot for trace elements contamination in Peninsular Malaysia *Bull. Environ. Contam.Toxicol.* **84**(5) 568-573.

[9] Mahat N A, Muktar N K, Ismail R, Abdul Razak F I, Abdul Wahab R, and Abdul Keyon A S 2018 Toxic metals in Pernaviridis mussel and surface seawater in Pasir Gudang coastal area, Malaysia, and its health implications. *Environ. Sci. Poll. Res.* **25**(30) 30224-30235.

[10] Yap C K, Peng S H and Leow C S 2019 Contamination in Pasir Gudang Area, Peninsular Malaysia: What can we learn from Kim Kim River chemical waste contamination? *J. of Humanities and Education Dev.*. **1** 82-87.

[11] Crosby N T, Day J, Hardcastle W A, Holcombe D G and Treble R D 1995 Quality in the analytical chemistry laboratory (New York: John Wiley & Son).

[12] Elias M S, Samuding K, Rahman S A and Wo, M Y 2018 Assesment of toxic elements in sediments of Linggi Rver using NAA and ICP-MS techniques. *MethodX* **5** 454-465.

[13] Al-Heety L F D, Hasan O M and Al-Heety E M S 2021 Assessment of heavy metal pollution of plants grown adjacent to power generators in Ramadi city. *Earth and Environ. Sci.* **779**(012023).

[14] Liu W X, Liu J W, Wu M Z, Li Y and Zha Y, Li S R 2009 Accumulation and translocation of toxic heavy metals in winter wheat (Triticum aestivum L.) growing in agricultural soil of Zhengzhou, China *Bull. Environ. Contam. Toxicol.* **82**(3) 343–347.

[15] DOE 2009. Contaminated land management and control guidelines no. 1: Malaysian recommended site screening levels for contaminated land https://www.doe.gov.my/wp-content/uploads/2021/09/1.-Contaminated-Land-Management-and-Control-Guidelines-No-1_Malaysian-Recommended-Site-Screening-Levels-for-Contaminated-Land.pdf.

[16] Kabata-Pendia 2001Trace Elements in Soils and Plants 3rd ed. (Boca Raton, FL: CRC Press).

[17] Doğanlar Z B, Atmaca M 2011 Influence of airborne pollution on Cd, Zn, Pb, Cu, and Al accumulation and physiological parameters of plant leaves in Antakya (Turkey). Water, *Air, & Soil Pollut.* **214** 509-523.

[18] Filipović-Trajković R, Ilić Z S, Šunić L and Andjelković S 2012 The potential of different plant species for heavy metals accumulation and distribution *J. of Food, Agri. & Environ.* **10** (1) 959-964.
[19] Gowariker V, Krishnamurthy V, Gowariker S, Dhanorkar M and Paranjape K 2009 *The Fertilizer Encyclo.* (John Wiley & Sons).
[20] FAO/WHO 2007 Summary of evaluations performed by the joint FAO/WHO expert committee on food additives (JECFA 1956-2007) (first through 68th meetings) Food and Agriculture Organization of the United Nations and the World Health Organization (Washington, DC., US: ILSI Press International Life Sciences Institute).
[21] Cakmak I, Öztürk L, Karanlik S, Marschner H, Ekiz H 1996 Zinc-efficient wild grasses enhance release of phytosiderophores under zinc deficiency *J. Plant Nutr.* **19** 551–563.
[22] Mousavi S R, Shahsavari M and Rezaei M 2011 A general overview Manganese (Mn) importance for crops production *Australian J. of Basic and Appl. Sci.* **5**(9): 1799-1803.
[23] Santiago A, Stefanie H, Bastian M, Edgar P 2020 Manganese in plants: from acquisition to subcellular allocation, Frontiers in Plant Sci. **11** 300 https://www.frontiersin.org/article/10.3389/fpls.2020.00300.
[24] Emamverdian A, Ding Y, Mokhberdoran F and Xie Y 2015 Heavy metal stress and some mechanisms of plant defense response. *Scientific World J.* **2015** 1-12.

Acknowledgements
The authors wish to thank the Department of Environment (DOE) for the sampling process in the Pasir Gudang industrial area. Special thanks to staff members of ACA and e-Tag for their relentless help and support during the undertaking of this project. The authors also would like to thank Dr Mazleha Maskin for her helpful advice on statistical analysis.

Assessment and sources identification of air quality pollution in Klang Valley, Kuala Lumpur, Malaysia

Md Suhaimi Elias[1,a)], Azian Hashim[1], Nurul Fairuz Dyana Bahrudin[2], Nur Aqilah Sapiee[3], Wilfred Paulus[3], Muhammad Azfar Azman[1], Nor Jawahir Raduian[4], and Intan Masayu Abdullah[d]

[1]Analytical Chemistry Application Group, Waste and Environmental Technology Division, Malaysian Nuclear Agency, 43000, Kajang, Selangor, Malaysia.
[2]Environmental Trace Application Group, Waste and Environmental Technology Division, Malaysian Nuclear Agency, 43000, Kajang, Selangor, Malaysia.
[3]Material Technology Group, Industrial Technology Division, Malaysian Nuclear Agency, 43000, Kajang, Selangor, Malaysia
[4]Conservation Management Division, Department of Museum Malaysia, Ministry of Culture and Tourism Malaysia, Damansara Road, 50566, Kuala Lumpur, Malaysia

[a]mdsuhaimi@nuclearmalaysia.gov.my

Abstract. Air quality pollution increases a research interest due to its importance in influencing weather and temperature, restricting visibility and causing harmful effects on human health. Samples of airborne particulate matter (APM) were collected weekly during the period from 2018 to 2020 at Klang Valley, Kuala Lumpur. The samples were collected using a Gent stacked filter sampler in two fractions particulate matter $PM_{2.5}$ ($PM_{2.5}$ – size of particulate < 2.5 μm) and particulate matter PM_{10} (PM_{10} – size of particulate 2.5 – 10 μm). 100 pairs of APM samples were analysed by Energy Dispersive X-Ray Fluorescence (EDXRF) for their elemental composition. The analysis result of air quality in Klang Valley, Kuala Lumpur, for an average mass concentration of $PM_{2.5}$ (26.8 μg/m^3) and PM_{10} (44.8 μg/m^3) were higher compared to the interim value of Malaysian ambient air quality standard 2020 (MAAQS-2020) ($PM_{2.5}$ = 15.0 μg/m^3; PM_{10} = 40.0μg/m^3). The results showed the possible sources of air pollution in the Klang Valley contributed by biomass burning/smoke (31.6%), soil dust (10.3%), industries (8.4%), motor vehicles (7.1%), sea spray (6.3%) and secondary sulphate (5.5%).

1. Introduction

Air quality is one of the major issues in many countries in the Asian region, including Malaysia. Air quality in Klang Valley, especially in the Kuala Lumpur area, has become a big concern to the government as this area is considered an entry point and tourism destination. Many efforts have been made to increase air quality by encouraging the public to use public transport, car-pooling and enforcing a new air quality standard guideline. The Klang Valley region is an area in Malaysia with rapid development and a high-density population[1]. The area comprises the capital city of Kuala Lumpur, with an estimated population of more than 5 million. Rapid growth with uncontrolled industrial development[2], biomass and fossil fuel combustion, increasing numbers of motor vehicles[3–7] and human activities such as land clearing, development of roads and highways and residential with a high growth population in Klang Valley are responsible for the emissions and release of various types of pollutants in the area[1].

The study of elemental composition in air filters collected from the Kuala Lumpur area is essential to identify the level and the sources of air pollution in Malaysia. The previous study of fine particulate matter 2.5 ($PM_{2.5}$) for the samples collected from 2000 to 2006 identified five sources of aerosols in the Klang Valley. These sources include sea spray, motor vehicles, smoke, industry, soil and an unknown source[8]. However, the study of air pollution collected from 2004 to 2008 found that the elemental composition in fine particles from the Kuala Lumpur (Klang Valley) area mainly originated from five anthropogenic sources. The sources of air pollution namely two-stroke engines (35.7%), motor vehicles (31.9%), smoke/biomass burning (17.5%), soil dust (8.3%) and industry (16.7%)[1]. The results show that 5 sources were identified in fine $PM_{2.5}$ filters collected from Serpong, Indonesia, namely lead industry mixed with road dust (12%), diesel vehicles (30%), oil and coal-fired power plant (26%), road dust (17%) and biomass burning mixed with road dust (15%)[5]. The fine $PM_{2.5}$ filters collected from Ulaanbaatar identified the seven source profiles. The air pollution sources of Ulaanbaatar originated from soil (crustal matter), coal combustion (black carbon with significant sulphur), coal combustion (less sulphur component), motor vehicles, biomass burning, road dust and zinc[7].

This study aims to identify the sources of air pollution and their fingerprints, contributing to the significant criteria for air quality in Klang Valley, Malaysia. This study focuses on monitoring and analysing $PM_{2.5}$, PM_{10}, and elemental content in air particulate samples and to assist the Department of Environmental (DOE), Malaysia, in developing MAAQS, especially for the standard of elemental content.

2. Methodology

2.1. Mass and black carbon measurement of air filter sample

The total mass of air filter samples was determined by weighing the filter using a microbalance (METTLER, Model MT5). The balance was equipped with a Po–212 (alpha emitter) electrostatic charge eliminator (STATICMASTER) to eliminate the static charge accumulated on the filters before each weighing. Black carbon measurements of air filters were performed using an EEL Smoke Stain Reflectometer, and the black carbon concentration was determined according to the method for 47 mm polycarbonate filter[9].

2.2. Sampling of air particulate matter (APM)

All air filter (particulate) samples were collected using a Gent stack sampler. The sampler was placed on a rooftop of a two-story building at the University Technology Malaysia (UTM), Kuala Lumpur, at approximately 10 meters in height (Figure 1). The coordinate of the sampling location at $3^{o}\,10'\,30''$ N, $101^{o}\,43'\,24''$ E just north of the city center park and the Petronas Tower, Kuala Lumpur. The sampler was programmed to run automatically at an airflow rate of 15 L/min to collect two fractions of particulate matter $PM_{2.5}$ ($PM_{2.5}$ – size of particulate matter < 2.5 μm) and particulate matter PM_{10} (PM_{10} – size of particulate matter 2.5 – 10 μm) for 8 hours at the cycle of 1 day (24 hours) with a frequency of 2 times/week.

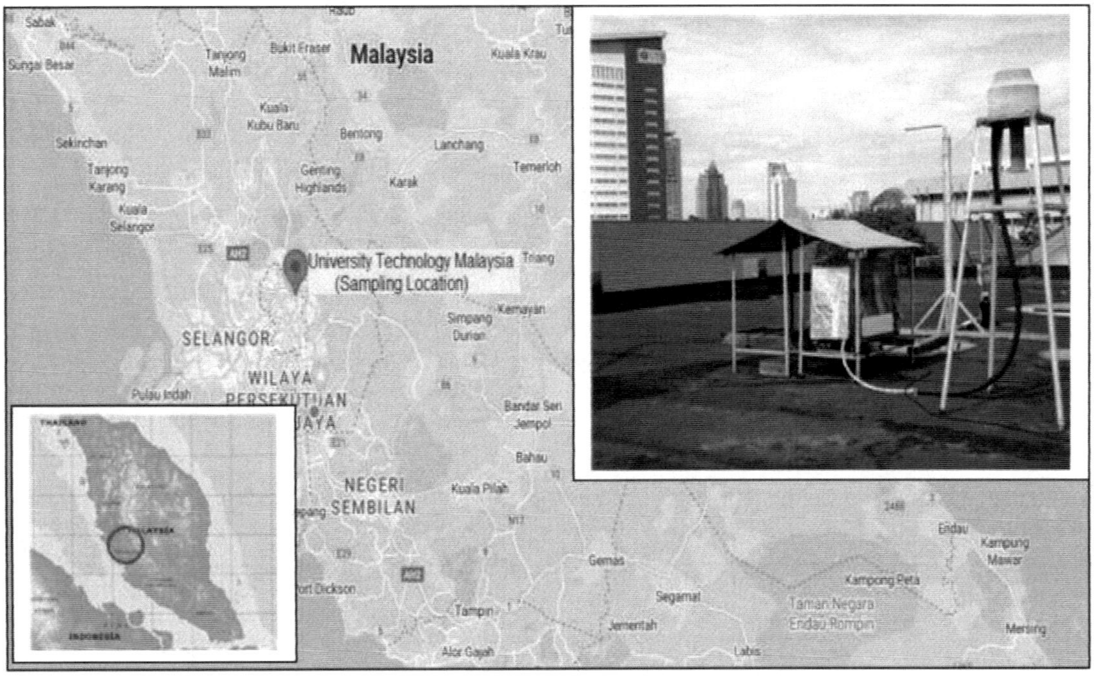

Figure 1. Sampling location of APM at two-story building UTM, Kuala Lumpur.

2.3. Analysis of elemental content of air filter by using XRF

A multifunction energy dispersive X-ray fluorescence (EDXRF) of the QUANT'X system from Thermo Scientific was employed to determine the concentration of major and trace elements in $PM_{2.5}$ filter samples. This EDXRF spectrometer using an air-cooled X-ray tube with Rh target. The tube has a power of a maximum of 50 watts. The exited X-rays from bombarded samples were detected by a Si (Li) detector with a resolution of 155 eV for 5.9 keV X-rays. The measurement was performed under the following conditions: X-ray generator voltage of 6 – 30kV and current of 0.84 – 1.90mA, measuring time 60-150 s, dead time 50%, under atmospheric air. All of the XRF spectra were quantitatively analysed with the use of the Thermo' Win Trace software version 4. The EDXRF analysis method was developed by applying a linear technique calibration curve to perform quantitative analysis. Standard reference material (NIST- SRM 2783) was analysed for quality assurance and quality control.

3. Results and Discussion

3.1. Analysis result of daily mass concentration for $PM_{2.5}$, PM_{10} and black carbon in air filter

Analysis results of mass concentration of $PM_{2.5}$, PM_{10} and black carbon (BC) in air filter samples were depicted in Figures 2(a) - (c), respectively. The previous air quality standard was set up in 2018 (interim value) for Malaysian air quality index are 45μg/m^3 (yearly) and 120 μg/m^3 (24 hour) for PM_{10} whilst 25 μg/m^3 (yearly) and 50 μg/m^3 (24 hours) for $PM_{2.5}$. Malaysia has set up a new standard for the air quality index. Nowadays, the air quality standard is referred to as the implementation of the Malaysian ambient air quality standard 2020 (MAAQS – 2020). The air quality standard for the interim value of Malaysian ambient air quality standard (MAAQS -2020) for PM_{10} is 40 μg/m^3 (yearly) and 100 μg/m^3 (24 hours), whilst interim values for $PM_{2.5}$ are 15 μg/m^3 (yearly) and 35 μg/m^3 (24 hours).

The result showed the mass concentration of $PM_{2.5}$ and PM_{10} from January 2018 to March 2020 was relatively showed a similar trend except for August to October 2019, which is the mass concentration of $PM_{2.5}$ and PM_{10} was slightly high (Figure 2(a) and (b)). The mass concentrations of $PM_{2.5}$ and PM_{10} increased due to the southwest monsoon dry season between May to October. The obtained result indicates the daily $PM_{2.5}$, PM_{10} and black carbon (BC) levels recorded between Jan 2018 to March 2020 were in the range of 2.9 – 71.7 μg/m^3, 7.1 – 111.7 μg/m^3, and 0.06 – 6.86 μg/m^3, respectively. The average concentration of $PM_{2.5}$ and PM_{10} for Jan 2018 to March 2020 was 26.8 μg/m^3 and 44.8 μg/m^3, respectively. This indicates the average concentration of $PM_{2.5}$ and PM_{10} for air quality in Klang Valley, Kuala Lumpur, is higher compared to the Malaysian ambient air quality standard 2020 (MAAQS-2020), as shown in Figure 2(a) and Figure 2(b), respectively. Summary statistical data analysis results of the mass concentration of $PM_{2.5}$, PM_{10} and black carbon (BC) in air filter samples were tabulated in Table 1.

Table 1. Summary statistics for measurement of $PM_{2.5}$, PM_{10} and Black Carbon (BC) in air filter samples.

	$PM_{2.5}$ (μg/m^3)	PM_{10} (μg/m^3)	Black Carbon (BC) (μg/m^3)
N	196	196	196
Min	2.86	7.14	0.06
Max	71.7	111.7	6.86
Mean	26.8	44.8	2.6
Median	25.5	43.8	2.7
Sum	5255	8784	502
Std. error	0.76	1.10	0.10
Variance	113.9	235.9	1.9
Stand. dev	10.7	15.4	1.4
25 percentil	18.9	33.8	1.4
75 percentil	32.9	53.8	3.6
Coeff. var	39.8	34.3	53.6

N = Number of samples

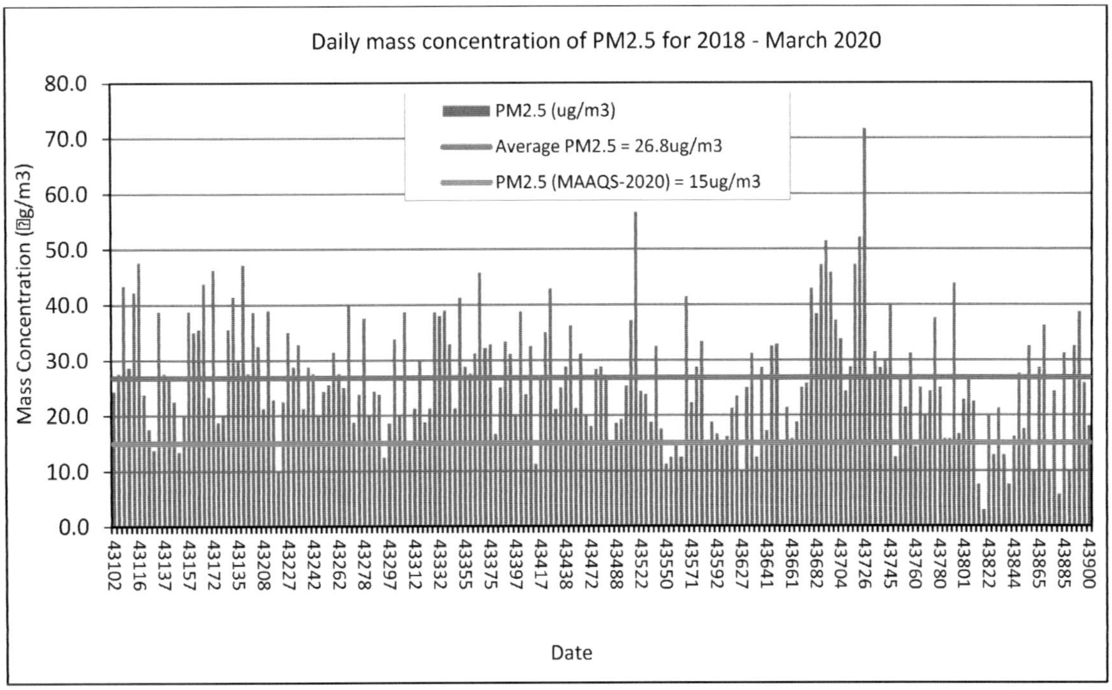

Figure 2(a). Analysis result of mass concentration of $PM_{2.5}$ in fine air filter

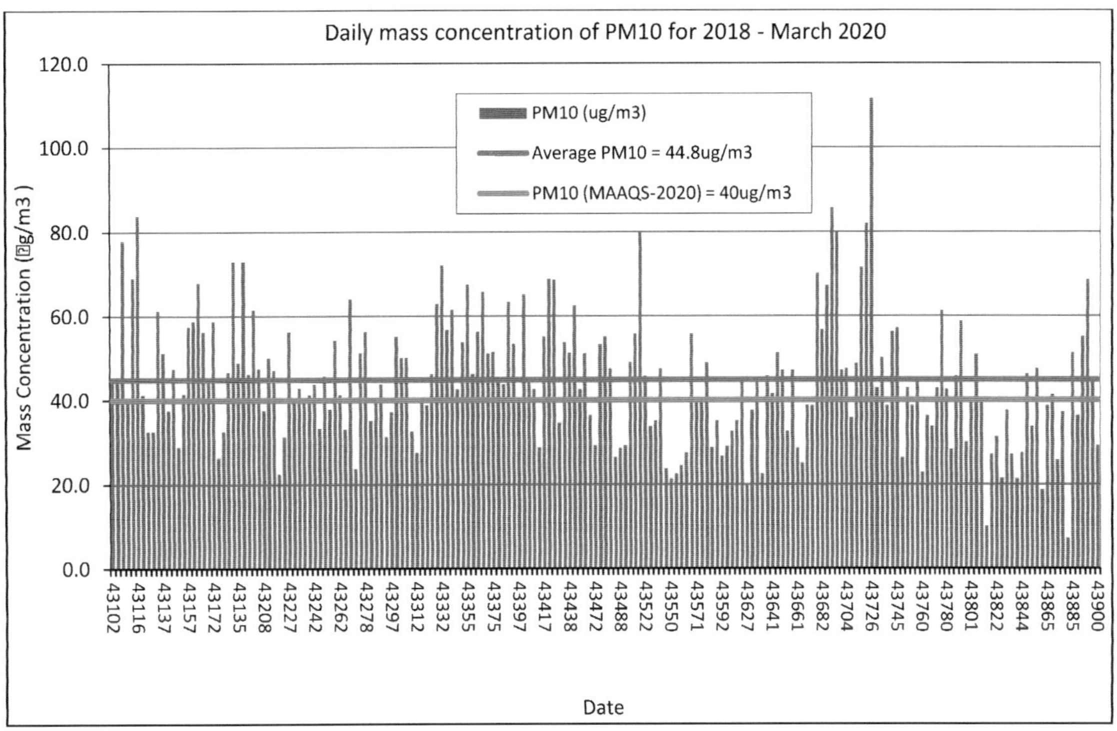

Figure 2(b). Analysis result of mass concentration of PM_{10} in coarse air filter

IOP Conf. Series: Materials Science and Engineering 1285 (2023) 012017 doi:10.1088/1757-899X/1285/1/012017

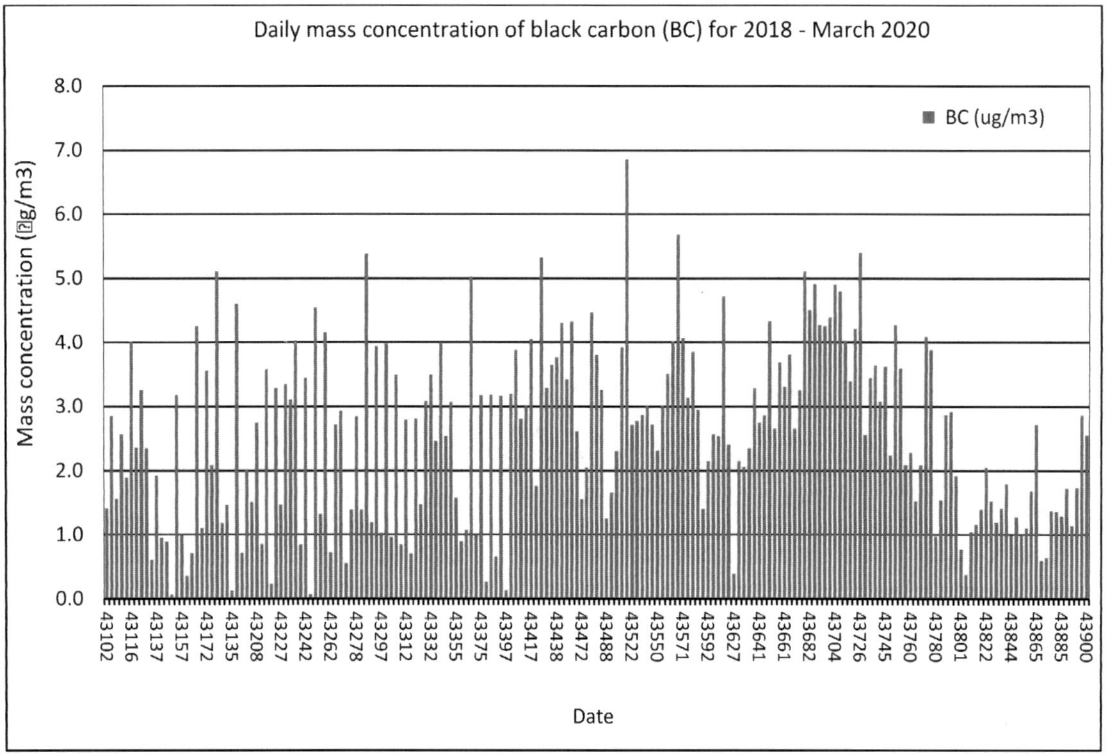

Figure 2(c). Analysis result of mass concentration of black carbon (BC) in fine air filter

3.2 Result of elemental content in air filter samples.

In this study, 18 elements were analysed by XRF for air filter samples collected from 2019 to March 2020. The air filter samples collected for 2018 were unable to be analysed for the elemental content due to physical defects (breakable, brittle). The bar graph of major (Na, Mg, Al, S, Si, K, Ca) and trace elements (Pb, Ti, V, Cr, Mn, Ba, Fe, Ni, Cu, Zn, As) concentration in air filter were depicted in Figure 3(a) – (r), respectively. The elemental content indicates the normal trend, where low concentrations were observed from November to February during the wet season. Higher concentrations were observed during the dry season between May to October for the elements of Mg, Al, S, Si, K, Ca, Pb, Ti, V, Mn, Fe, Ni, Cu, Za and As except for the element of Na, Cr and Ba. Generally, uncontrolled development of residential, metal and electronics industries, biomass burning, vehicles and human activities are responsible for the emissions of various pollutants in the Klang Valley area[1]. Summary of statistics for analysis of concentration in air filter and analysis result of air particulate on filter media (SRM 2783) of major and trace elements were tabulated in Table 2.

IOP Conf. Series: Materials Science and Engineering 1285 (2023) 012017 doi:10.1088/1757-899X/1285/1/012017

Table 2. Analysis of elemental concentration in fine air filter and SRM 2783 for major and trace elements.

	N	Min	Max	Mean	Median	Certified value of SRM 2783 (ng)	Uncertainty of certified value of SRM 2783	Measured value of SRM 2783 (ng)	Uncertainty of measured value of SRM 2783	U-Test score
Al	100	16.0	1494	280.0	138.0	23210	530	27350	1560	2.51
As	100	1.0	4.0	1.6	1.0	11.8	1.2	11.0	1.0	0.43
Ba	100	3.0	37.0	17.5	17.5	335	50.0	327.5	63.0	0.09
Ca	100	89.0	782.0	275.7	252.0	13200	1700	14030	2620	0.27
Cr	100	1.0	8.0	3.6	3.0	135	25.0	129	25.0	0.17
Cu	100	1.0	92.0	25.7	25.0	404	42.0	475.5	36.0	1.29
Fe	100	56.0	466.0	177.6	172.0	26500	1600	25980	354	0.32
K	100	20.0	1359	493.8	461.0	5280	520	6088	24.0	1.55
Mg	100	5.0	252	86.4	85.0	8620	520	7440	1358	0.81
Mn	100	0.0	21.0	7.2	7.0	320	12.0	378	20.0	2.49
Na	100	35.0	1713	413.4	298.5	1860	100	1475	361	1.03
Ni	100	1.0	24.0	4.8	4.0	68	12.0	51.5	11.0	1.03
Pb	100	2.0	77.0	27.8	26.0	317	54.0	335	21.0	0.31
S	100	184.0	2413	987.9	960.5	1050	260	1163	37.0	0.43
Si	100	60.0	1017	394.8	339.5	58600	1600	53500	1820	2.10
Ti	100	3.0	186.0	60.8	60.5	1490	240	1796	146	1.09
V	100	1.0	48.0	9.9	7.0	48.5	6.0	43.5	5.0	0.64
Zn	100	2.0	86.0	25.7	16.5	1790	130	1914	492	0.24

The calculation of the U-Test score was based on equation (1) in the reference[10].

$$u = \frac{|x - X_A|}{\sqrt{(\sigma_x)^2 + (\sigma_A)^2}} \tag{1}$$

where:

x : is measured value of standard reference material (SRM)
X_A : is certified value of standard reference material (SRM)
σ_x : is uncertainty of measured value of standard reference material (SRM)
σ_A : is uncertainty of certified value of standard reference material (SRM)

The following decision limits for the u-scores were established[10]:
$u \leq 1.64$ - reported result does not differ from the assigned value
$1.64 < u \leq 1.95$ - reported result probably does not differ from the assigned value
$1.95 < u \leq 2.58$ - it is not clear whether the reported and assigned values differ
$2.58 < u \leq 3.29$ - reported result is probably different from the assigned value
$3.29 < u$ - reported result differs from the assigned value

The result of the U-Test score for the analysis of SRM showed less than 3 indicating the result of analysis for the air filters (air particulate) samples is considered acceptable. The results of the U-Test score for the analysis of SRM ranged from 0.09 – 2.51 (Table 2).

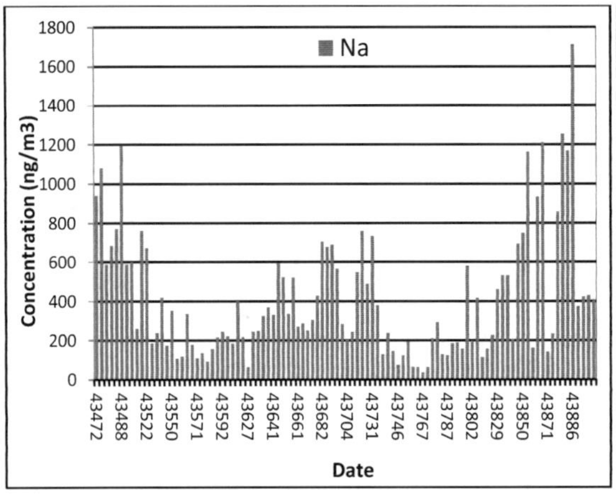

Figure 3(a). The bar graph of Na concentration in air filter.

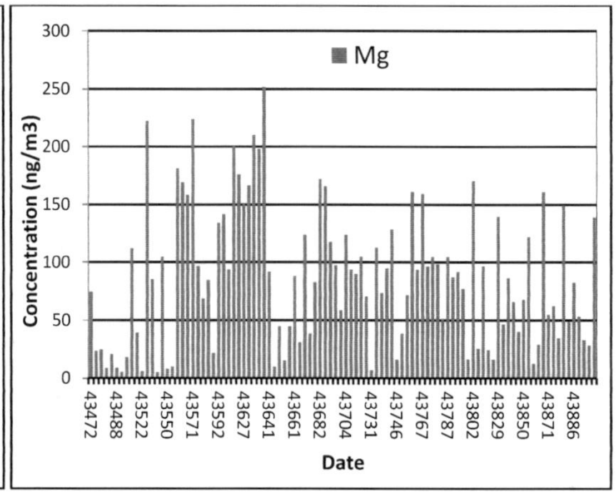

Figure 3(b). The bar graph of Mg concentration in air filter.

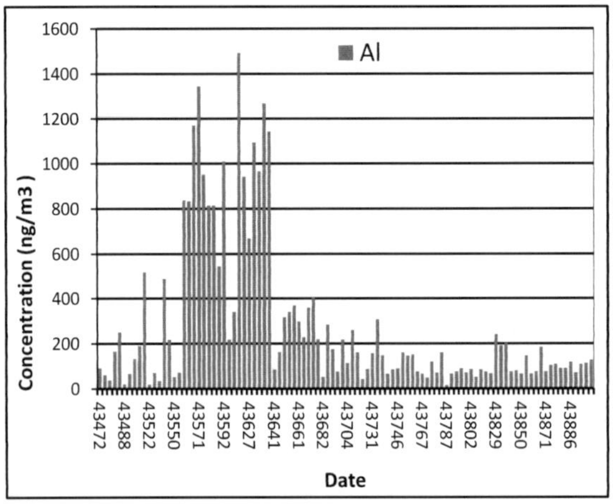

Figure 3(c). The bar graph of Al concentration in air filter.

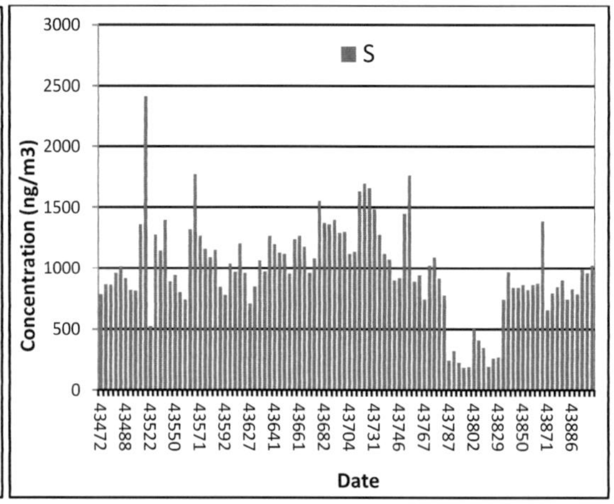

Figure 3(d). The bar graph of S concentration in air filter.

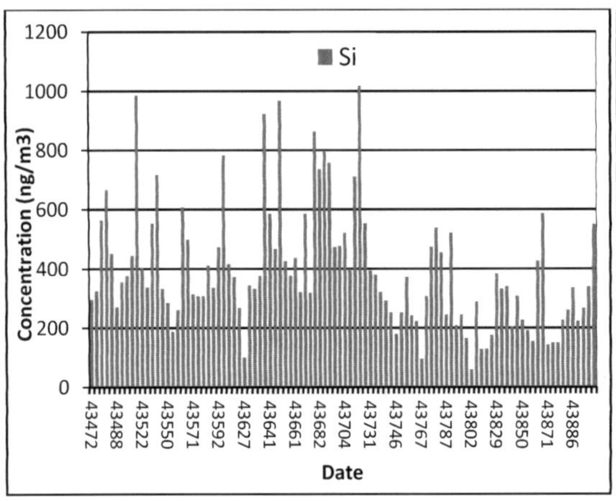

Figure 3(e). The bar graph of Si concentration in air filter.

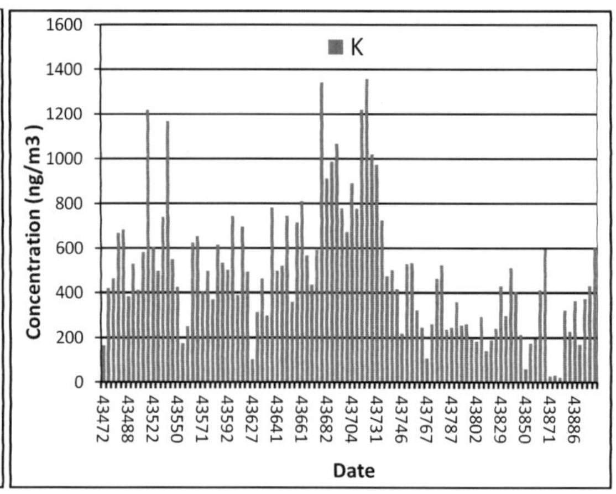

Figure 3(f). The bar graph of K concentration in air filter.

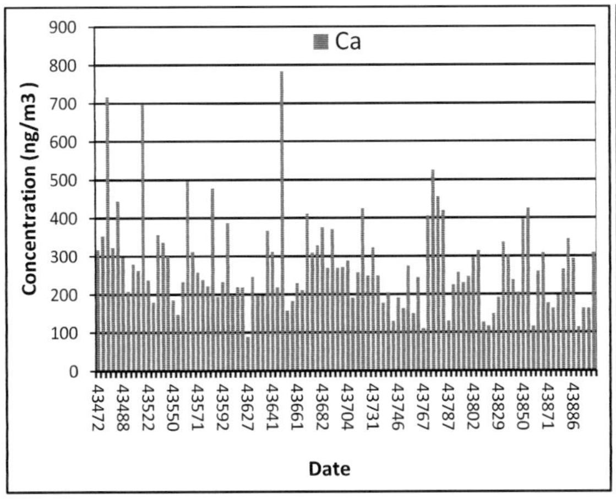

Figure 3(g). The bar graph of Ca concentration in air filter.

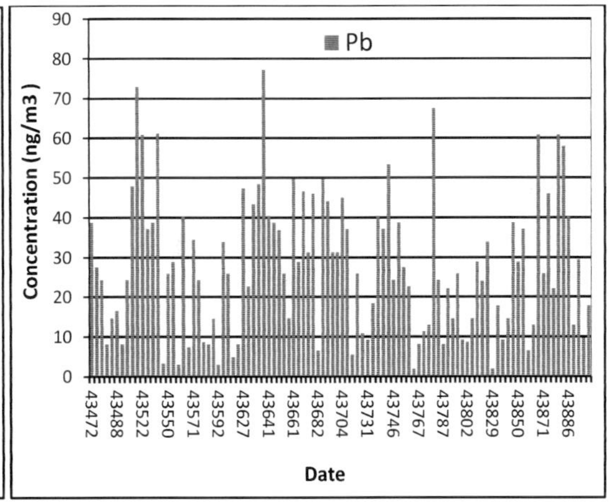

Figure 3(h). The bar graph of Pb concentration in air filter.

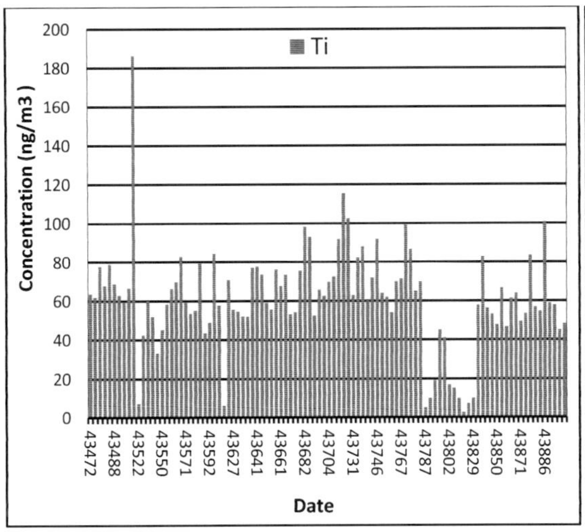

Figure 3(i). The bar graph of Ti concentration in air filter

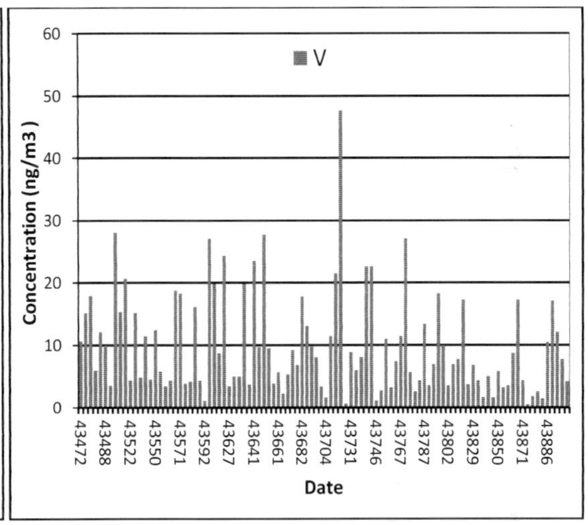

Figure 3(j). The bar graph of V concentration in air filter

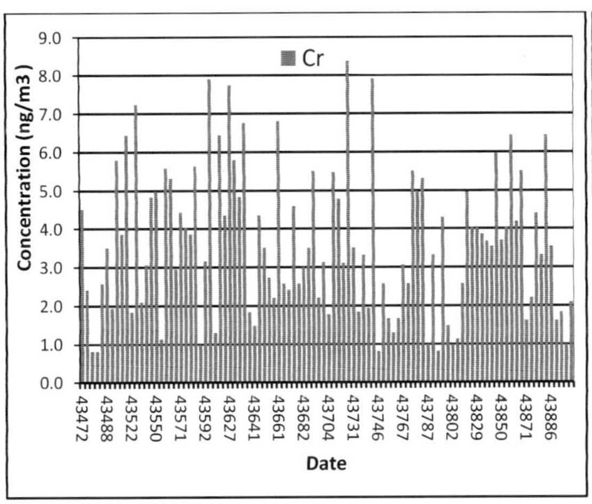

Figure 3(k). The bar graph of Cr concentration in air filter

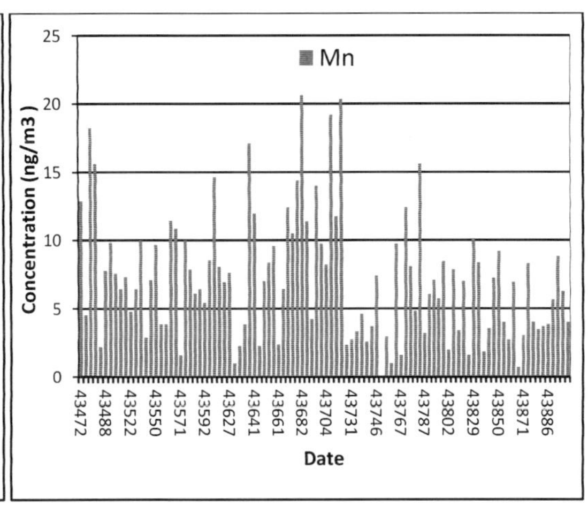

Figure 3(l). The bar graph of Mn concentration in air filter

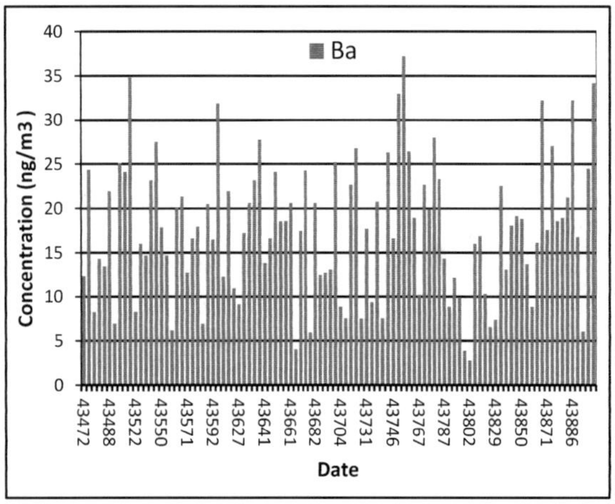

Figure 3(m). The bar graph of Ba concentration in air filter

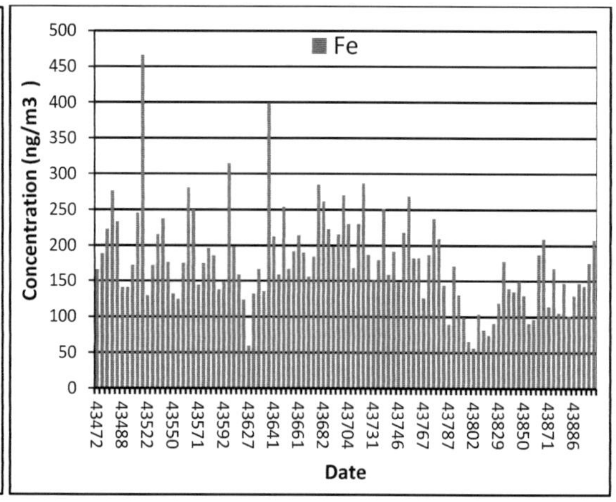

Figure 3(n). The bar graph of Fe concentration in air filter

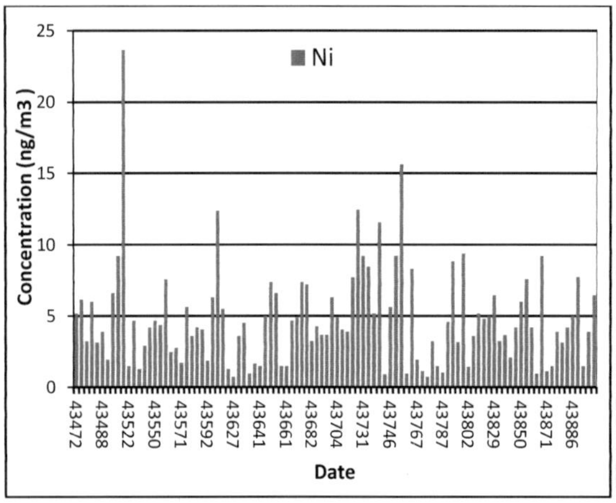

Figure 3(o). The bar graph of Ni concentration in air filter

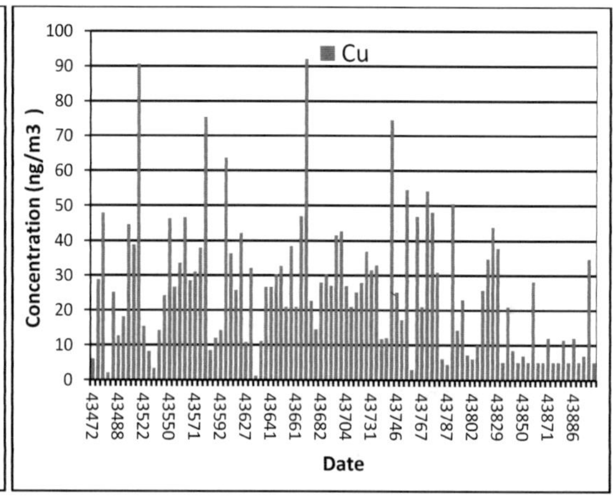

Figure 3(p). The bar graph of Cu concentration in air filter

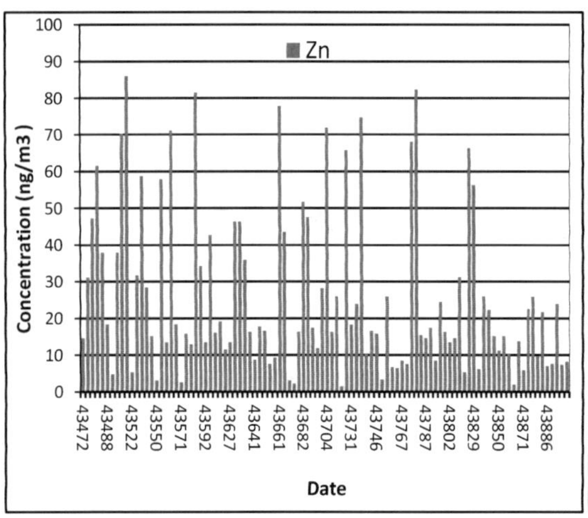

Figure 3(q). The bar graph of Zn concentration in air filter.

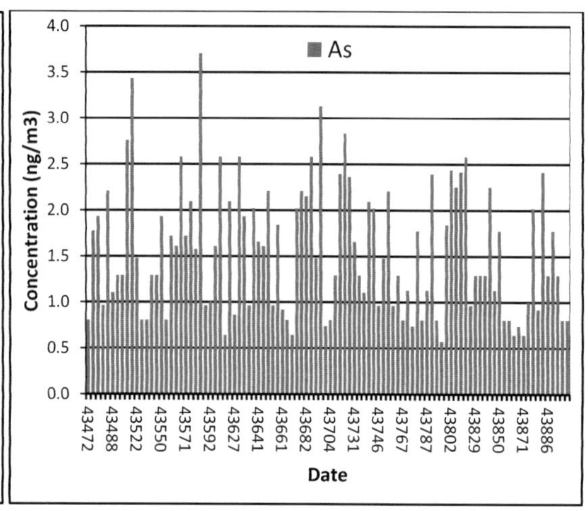

Figure 3(r). The bar graph of As concentration in air filter.

Figure 3(a) – (r). The bar graph of major and trace elements concentration in the fine air filter.

3.3. Back trajectory and long-range transport of smoke

Forest fires are common in Malaysia and Indonesia during dry months of the dry season, which runs from March to October, despite Malaysia and Indonesia having efforts to crack down on illegal burning. Sometimes fires are the result of slash and burn deforestation to clear land for farming or other agricultural activities.

Analysis results of the $PM_{2.5}$ graph showed anomalies in August and Sept 2019. The high concentration of daily mass of $PM_{2.5}$ showed on 20 August 2019 and 17 September 2019, as depicted in Figure 4(a). The time series plot for the K (smoke) concentration of Klang Valley from 2019 to March 2020 is shown in Figure 4(b). A similar trend was observed for the plotting graph of the K element, where the K element concentration showed higher on 20 August 2019 and 17 September 2019 (Figure 4(b)). It can be seen that there are two sharp peaks that occur on 20 August 2019 and 17 September 2019, where the K values were 1399ng/m^3 and 1698 ng/m^3, respectively.

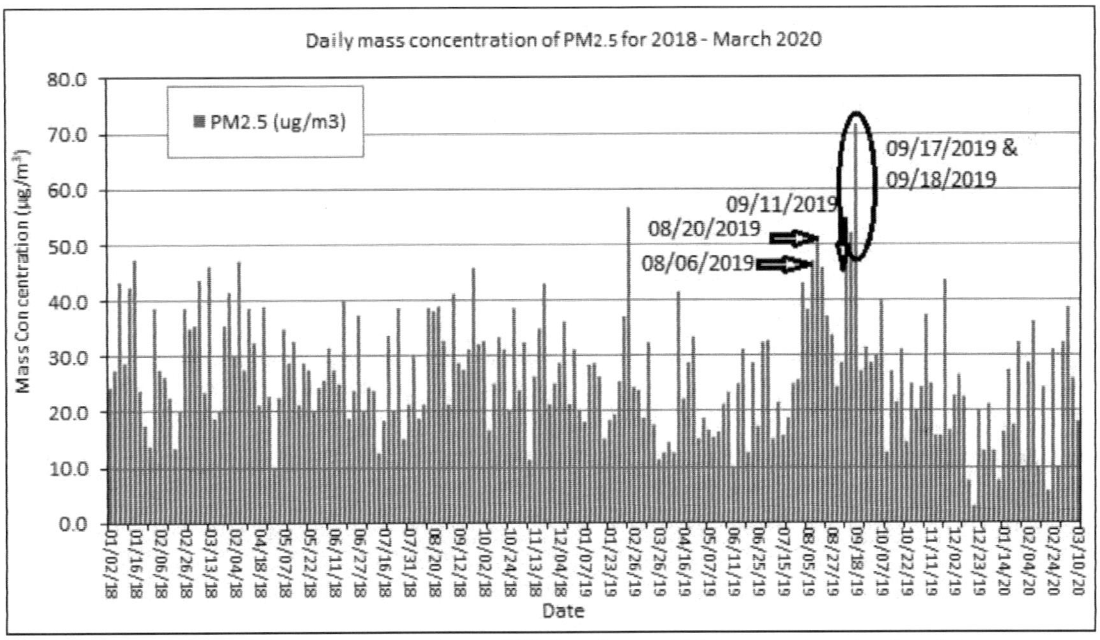

Figure 4(a). Mass concentration of $PM_{2.5}$ showed higher at the stipulated date.

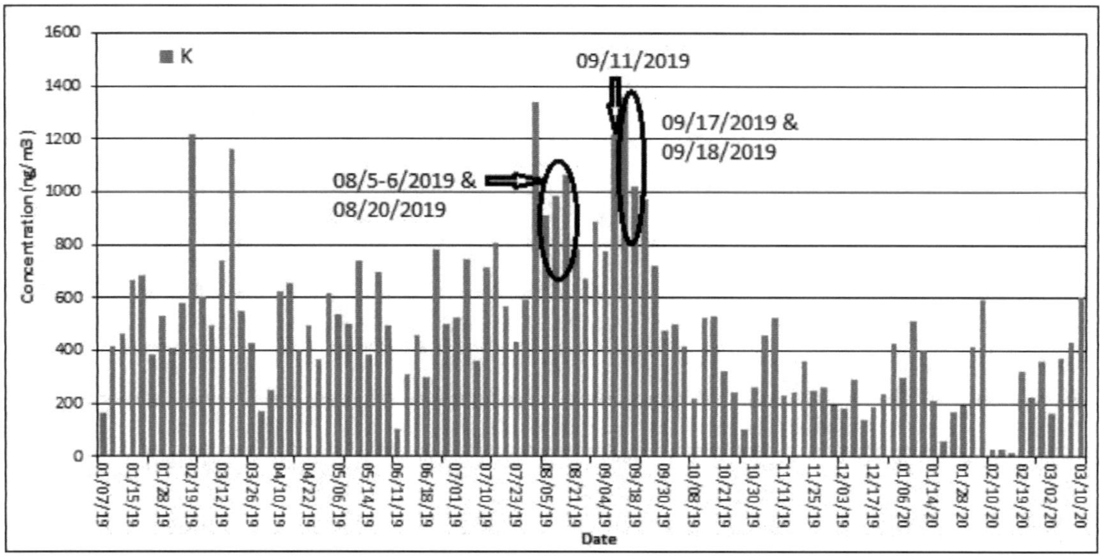

Figure 4(b). Concentration of potassium (K) showed higher at the stipulated date.

Smoke transboundary events for the selected studying period in Malaysia (Klang Valley) were identified based on the K element. From May to October, the south-westerly wind is blowing from Sumatra to Malaysia and bringing along thick smoke, blanketing part of Malaysia, Singapore and Indonesia for weeks.

Back trajectory plots at different heights 100m, 500m, 1000m and 2000m above ground level (AGL) were used to calculate the four days backward trajectories. Four days back trajectories calculated for 06 August 2019 and 20 August 2019 are shown in Figures 5(a) and 5(b), respectively. Figure 6 shows the hot spots (fire - red dot) in Sumatra, Indonesia, on 20 August 2019. The back trajectories calculated for 11 September 2019 and 17 September 2019 were shown in Figures 7(a) and 7(b), respectively, whilst the hot spot (fire -red dot) in Sumatra, Indonesia, on 17 September 2019 was shown in Figure 8. Both events on 20 August and 17 September 2019 (Figure 6 and Figure 8) were seen to be associated with international transport of smoke from the area of Sumatra Island. The smoke signature travels about 300 km to 900 km in the southwest direction from the source location to reach the receptor site in Kuala Lumpur.

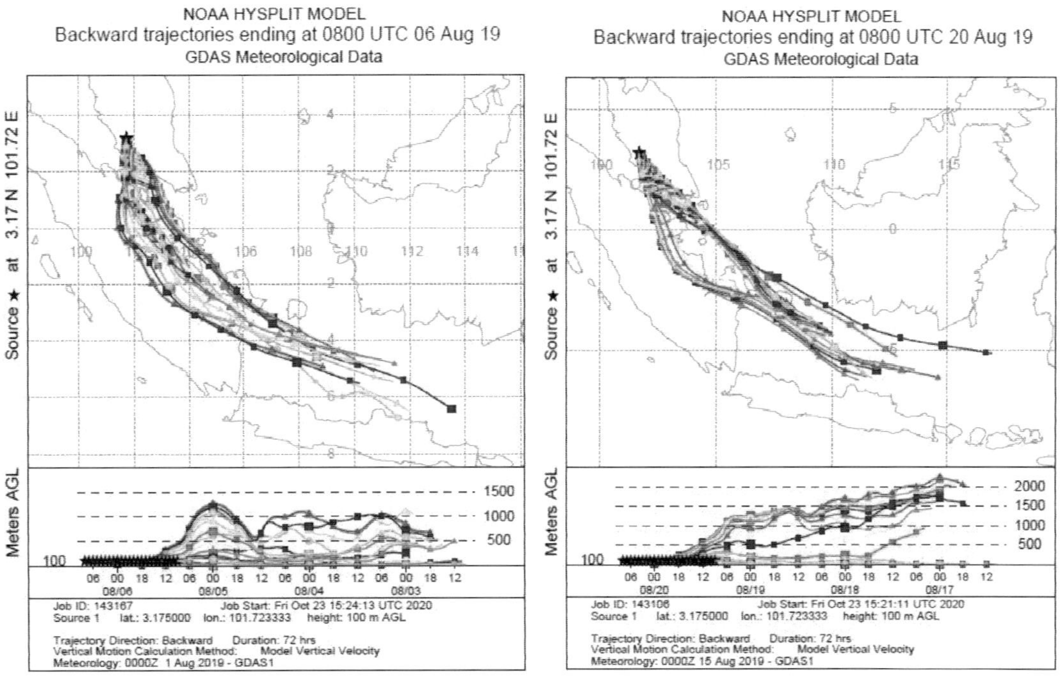

(a) (b)

Figure 5(a) – (b). Four days back trajectories were calculated for 06 August 2019 and 20 August 2019, respectively.

Figure 6. Hot spot (fire -red dot) in Sumatra, Indonesia, on 20 August 2019.

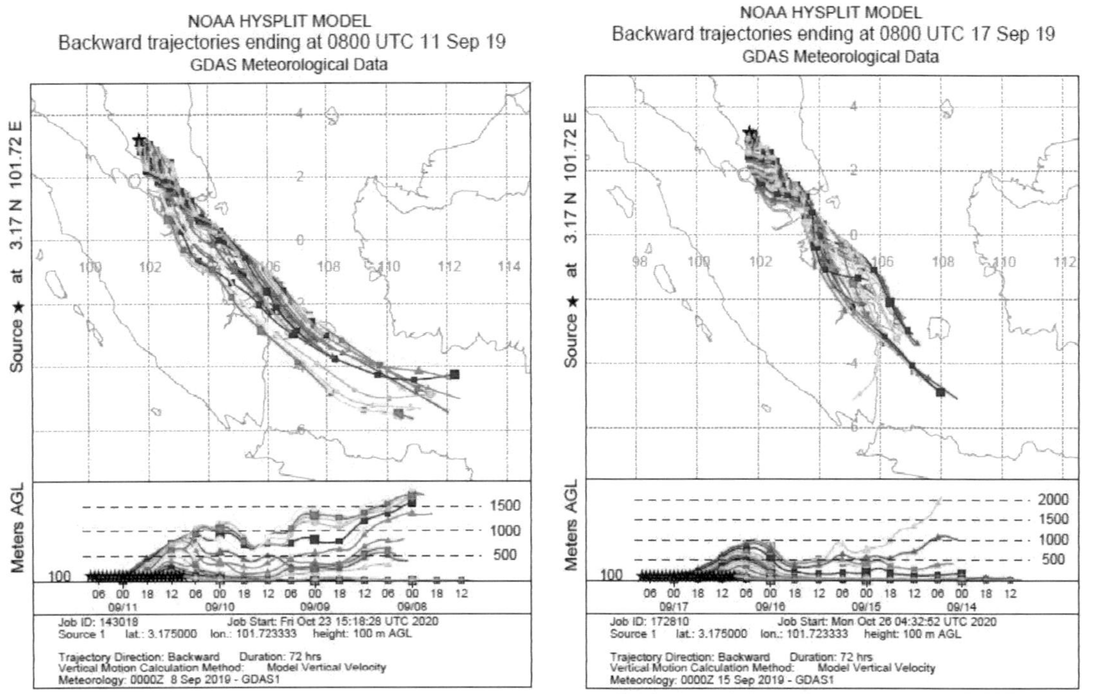

(a) (b)

Figure 7(a) – (b). Four days back trajectories were calculated for 11 September 2019 and 17 September 2019, respectively.

Figure 8. Hot spot (fire -red dot) in Sumatra, Indonesia on 17 September 2019.

3.4. Sources identification and fingerprint of pollution in fine air particulate matter.

The source identification for the elemental composition of fine aerosol in Klang Valley, Kuala Lumpur, was found to be a six-factor solution. Figure 9(a) – (f) presents the source identification profiles of air pollution in the Kuala Lumpur area.

The first factor, represents a smoke or biomass burning source that contributed about 31.6% of the fine mass concentration (Figure 9(a)). The sources fingerprint was characteristically high in black carbon (BC), K, Si, S, Fe, Ti, and Ca. Fine K is an accepted key element for smoke from biomass burning, which is released especially under high temperatures in fires mostly as potassium chloride and potassium sulphate [11,12].

The second factor is attributed to soil dust that contains high loading of Mg, Al, and Cr. The source contributed about 10.3% to the fine mass. The soil dust contains characteristic elements of Al, Si, Mg, K, Ca and Fe[13]. The characteristic of elements for soil dust in Klang Valley indicates high contains of Mg, Al, and Cr however, low contribution of Si, S and Fe as depicted in Figure 9(b).

The third factor has high loading of As, V, Ni, Cu, Cr, Ca, and Ti as shown in Figure 9(c). The source that contributed 8.4% of the fine mass was attributed to industries. The west coast of the Malaysian peninsula is an industrial area. Significant numbers of industries are involved in metal works scattered around Klang Valley and the main industrial area (covering Kapar, Shah Alam and Petaling Jaya) located about 15 to 20 km to the southern/southwestern side of the sampling site. The industries include iron and steel, paint, aluminium, fabrication, pharmaceutical, rubber glove production, and food industries.

The fourth factor represents a source from a motor vehicle that contributed about 7.1% of the fine mass (Figure 9(d)). The source of the fingerprint contained a characteristic element of Zn, Cr, Pb and Ni. The Kuala Lumpur city centre has approximately 1.5 million residents with high vehicles and is heavily affected by vehicle missions[1,14]. The sampling site lies between two highways to the north and south, surrounded by heavy traffic roads. One is 700 m in the north, and the other is 1 km south, and there are also busy roads to the west of the sampling site linked to Kuala Lumpur city centre. The higher loading of Zn is most likely related to motor vehicle emission[15] and also attributed to vehicle tires which contain the Zn compound used in tire manufacturing[16]. Ni mainly originated from petroleum and petroleum products[17]. Ni could remain as a residue in petrol and diesel and is mainly contributed by the incomplete combustion of petrol and diesel. The element of Cr could be from bearing and car break lining. The Pb could be produced from two-stroke engines and additional Pb in petrol as antiknock [17].

The fifth factor contributed about 6.3% of the fine mass concentration. The factor has high loading of Na, followed by Ca and Pb, which showed half of the fraction values compared to the element of Na, as shown in Figure 9(e). Besides that, the presence of trace elements of Si, Cr, Ba, V, Ti and Mn in the factor. The source for the fifth factor is mainly attributed to sea spray. The high Na and Cl loading suggests sea spray emission [1].

The last factor may be attributed to secondary sulphate. The factor contributed about 5.5% of the fine mass concentration and has a characteristic of high loading of Ba and S (Figure 9(f)). The presence of S indicates highly the possible source originated from secondary sulphate. The presence of Ti, Ni, Fe, V, Pb, Na, and Cu in the factor could be connected with the possible sources contributed by industries.

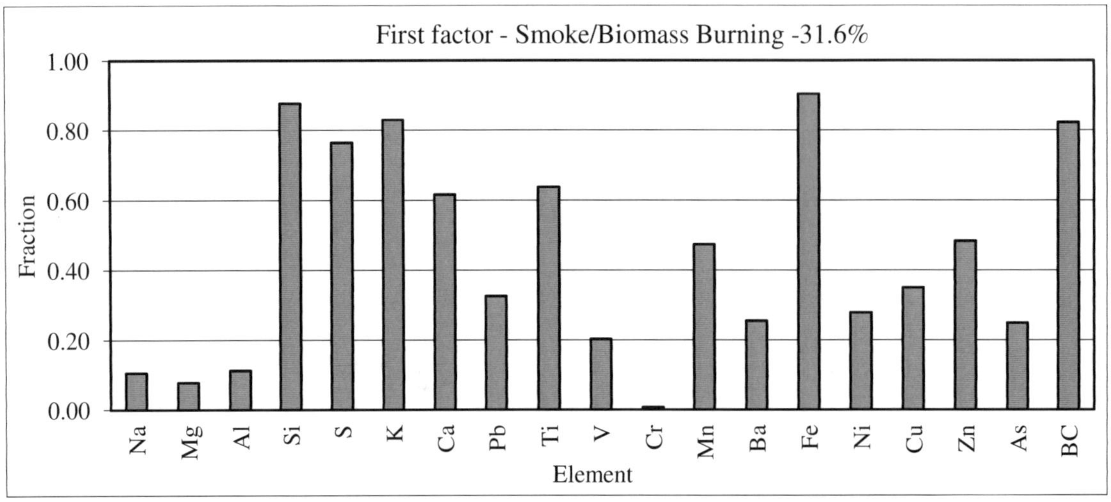
First factor - Smoke/Biomass Burning -31.6%
1.00
0.80
0.60
0.40
0.20
0.00
Fraction
Na
Mg
Al
Si
S
K
Ca
Pb
Ti
V
Cr
Mn
Ba
Fe
Ni
Cu
Zn
As
BC
Element

(a)

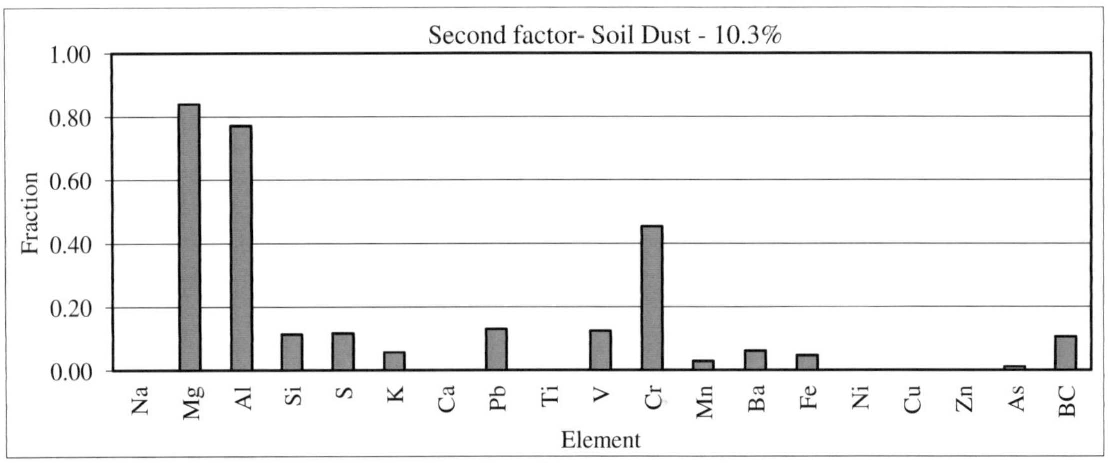
Second factor- Soil Dust - 10.3%
1.00
0.80
0.60
0.40
0.20
0.00
Fraction
Na
Mg
Al
Si
S
K
Ca
Pb
Ti
V
Cr
Mn
Ba
Fe
Ni
Cu
Zn
As
BC
Element

(b)

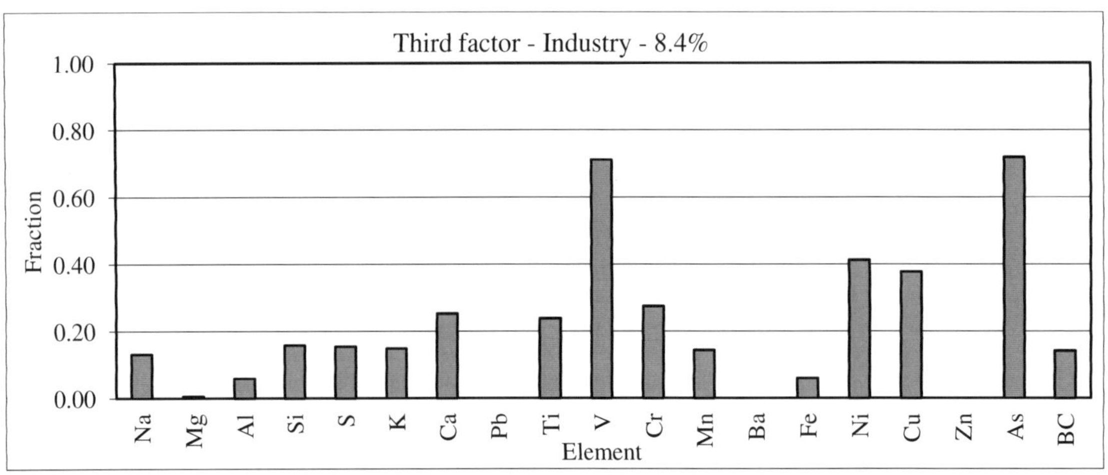
Third factor - Industry - 8.4%
1.00
0.80
0.60
0.40
0.20
0.00
Fraction
Na
Mg
Al
Si
S
K
Ca
Pb
Ti
V
Cr
Mn
Ba
Fe
Ni
Cu
Zn
As
BC
Element

(c)

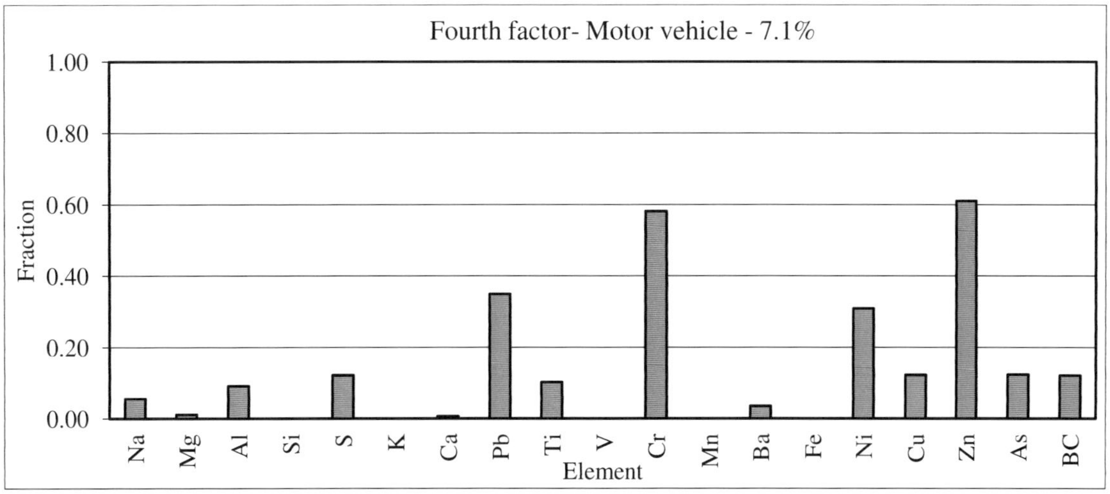

(d)

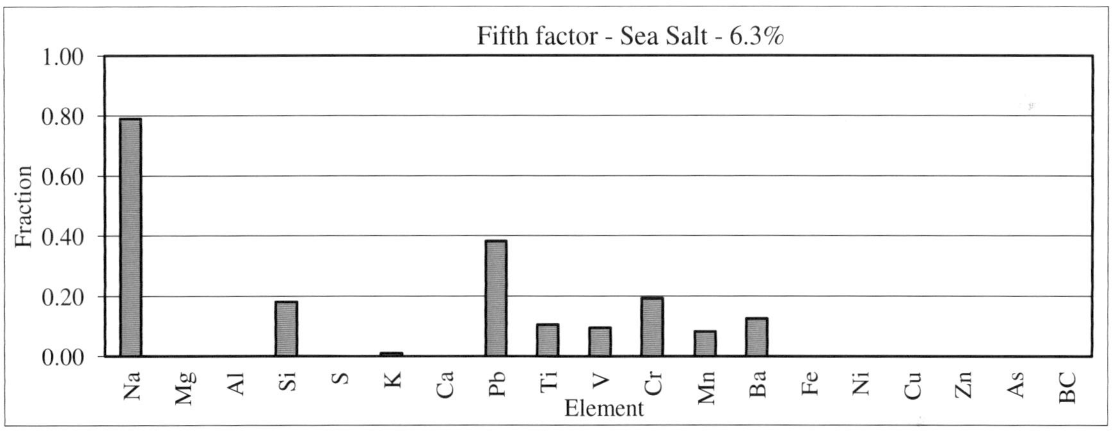

(e)

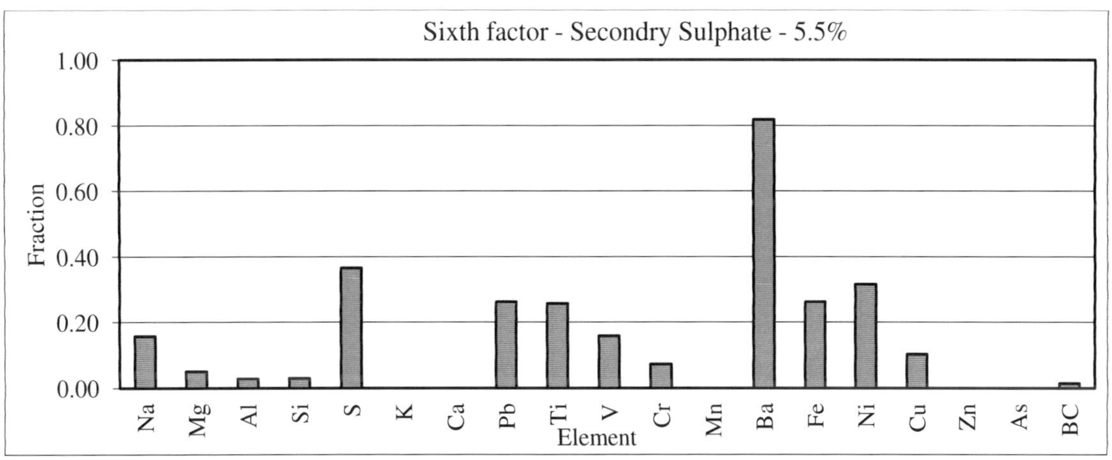

(f)

Figure 9(a) – (f). Source apportionment for the $PM_{2.5}$ mass measured at Kuala Lumpur, Klang Valley for 2019 - March 2020.

4. Conclusion

The obtained result indicates the average daily mass concentration of $PM_{2.5}$ (average = 26.8 μg/m^3) for the air particulate matter samples collected from Jan 2018 to March 2020 was higher than the Malaysian ambient air quality standard 2020 (MAAQS-20200) (average = 15 μg/m^3) whilst, the average daily mass concentration of PM_{10} (average = 44.8 μg/m^3) also showed slightly higher than MAAQS – 2020 (average = 40.0 μg/m^3). The source apportionment of air pollution in Kuala Lumpur, Klang Valley originated from smoke or biomass burning, soil dust, industries, motor vehicle, sea spray and secondary sulphate with the contribution of 31.6%, 10.3%, 8.4%, 7.1%, 6.3% and 5.5%, respectively. The back trajectory result indicates the source of smoke originated from the neighbouring country.

5. References

[1] Abdul Rahman S, Suhaimi Hamzah M, Wood A K, Elias M S, Abdullah Salim N A and Sanuri E 2011 Sources apportionment of fine and coarse aerosol in Klang Valley, Kuala Lumpur using positive matrix factorization *Atmos. Pollut. Res.* **2** 197–206

[2] Chan Y C, Simpson R W, Mctainsh G H, Vowles P D, Cohen D D and Bailey G M 1999 Source apportionment of PM2.5 and PM10 in Brisbane (Australia) by receptor modelling *Atmos. Environ.* **33** 3251–68

[3] Gulson B, Korsch M, Dickson B, Cohen D, Mizon K and Michael Davis J 2007 Comparison of lead isotopes with source apportionment models, including SOM, for air particulates *Sci. Total Environ.* **381** 169–79

[4] Cohen D D, Stelcer E, Hawas O and Garton D 2004 IBA methods for characterisation of fine particulate atmospheric pollution: A local, regional and global research problem *Nucl. Instruments Methods Phys. Res. Sect. B Beam Interact. with Mater. Atoms* **219–220** 145–52

[5] Santoso M, Lestiani D D, Mukhtar R, Hamonangan E, Syafrul H, Markwitz A and Hopke P K 2011 Atmospheric Pollution Research Preliminary study of the sources of ambient air pollution in Serpong , Indonesia *Atmos. Pollut. Res.* **2** 190–6

[6] Rahman S A, Elias S, Ashifa N, Salim A, Laili Z, Hashim A, Shukor S A and Omar S A 2018 Sources Apportionment of Fine and Coarse in Bangi, Selangor using Positive matrix Factorization *J. Sains Nukl. Malaysia* **30** 19–29

[7] Davy P K, Gunchin G, Markwitz A, Trompetter W J, Barry B J, Shagjjamba D and Lodoysamba S 2011 Air particulate matter pollution in Ulaanbaatar, Mongolia: Determination of composition, source contributions and source locations *Atmos. Pollut. Res.* **2** 126–37

[8] Rahman S A, Hamzah M S, Wood A K, Elias M S, Salim N A A and Sanuri E 2009 The application of factor analysis to identified source in klang valley aerosol *J. Nucl. Relat. Technol.* **6** 279–86

[9] Cohen, D.D., Taha, G., Stelcer, E., Garton, D., Box, G. 2000 Measurement and sources of fine particle elemental carbon at several key sites in NSW over the past eight years. *Proc. 15th Int. Clean air Conf. Sydney, Aust.* **27-30 Nov** 485–90

[10] PTNAAIAEA17 2019 Determination of Major, Minor and Trace Elements in a Land-plant material and a Siliceous material *Worldw. Open Profic. Test Neutron Act. Anal. Lab.* 1–152

[11] Baxla S P, Roy A A, Gupta T, Tripathi S N and Bandyopadhyaya R 2009 Analysis of diurnal and seasonal variation of submicron outdoor aerosol mass and size distribution in a northern indian city and its correlation to black carbon *Aerosol Air Qual. Res.* **9** 458–69

[12] Begum B A, Kim E, Biswas S K and Hopke P K 2004 Investigation of sources of

atmospheric aerosol at urban and semi-urban areas in Bangladesh *Atmos. Environ.* **38** 3025–38
[13] Watson J G, Chow J C and Houck J E 2001 PM2.5 chemical source profiles for vehicle exhaust, vegetative burning, geological material, and coal burning in Northwestern Colorado during 1995 *Chemosphere* **43** 1141–51
[14] Ebihara M, Chung Y S, Dung H M, Moon J H, Ni B F, Otoshi T, Oura Y, Santos F L, Sasajima F, Sutisna, Wee B S, Wimolwattanapun W and Wood A K B H 2008 Application of NAA to air particulate matter collected at thirteen sampling sites in eight Asian countries: A collaborative study *J. Radioanal. Nucl. Chem.* **278** 463–7
[15] Wei B, Jiang F, Li X and Mu S 2009 Spatial distribution and contamination assessment of heavy metals in urban road dusts from Urumqi, NW China *Microchem. J.* **93** 147–52
[16] Siong W B and Wood A K H 2006 Determination of Trace and Toxic Elements in Marine Sediments collected from the Strait of Malacca, Malaysia. *FNCA 2006 Work. Util. Res. React. Ctry. Rep.* 1–7
[17] Kabata-Pendias A and Szteke B 2015 Trace Elements in Abiotic and Biotic Environments *CRC Press. Taylor Fr. Group. Broken Sound Parkw. NW* 1–393

Acknowledgement

The authors would like to thank the Malaysian Nuclear Agency and the Regional Cooperative Agreement Regional Office (RCARO) for financial support under the research grant (RCARP1/RC03). We would also like to express our gratitude and special thanks to the Analytical Chemistry Application Group staff for their assistance in the sampling and analysis of samples.

Potential of tropical expansive clays as natural sealing layer in nuclear waste repository system: a preliminary study

A S N Shah[1,2,a)] and M A Zali[3]

[1]Materials Technology Group, Industrial Technology Division, Malaysian Nuclear Agency, 43000 Kajang, Selangor, Malaysia
[2]Department of Earth Sciences and Environment, Faculty of Science and Technology, National University of Malaysia, 43600 Bangi, Selangor, Malaysia
[3]Analytical Chemistry Application Group, Waste Technology & Environmental Division, Malaysian Nuclear Agency, 43000 Kajang, Selangor, Malaysia

[a]azlanshah@nm.gov.my

Abstract. The issue of radioactive waste disposal in Malaysia is frequently debated, particularly in terms of human and environmental long-term well-being. The creation of a repository with minimal hazard risk and cost implications is a major environmental concern related to nuclear waste. Expansive clay has distinct characteristics that are readily available in the environment and can serve as a natural barrier (sealing system) against the migration of radioactive and chemical pollutants. In this study, the physico-chemical and microstructural properties of 2 types of tropical soils with dominant clays mineral (vermiculite and montmorillonite) were investigated to better understand the potential of these clays as natural sealing in nuclear waste repository systems. Based on preliminary findings, both types of soil have been classified as alkaline soil (>7) with high swelling capacity (FSI) (>80%) and high consistency (LL>50%; PI>30%). The microstructural study demonstrates that vermiculite clays have a noticeable flaky morphology and a porous structure as compared to montmorillonite, indicating that the soil can absorb more water, hence decreasing its infiltration capability. Pinhole dispersion analysis reveals that montmorillonite clay is more prone to dispersion due to the presence of sodium ions and has a higher erosion potential than vermiculite clay. Future Malaysian stakeholders may employ this research to establish a long-term plan for the management of radioactive waste through the usage of natural resources.
Keywords. tropical expansive clay; nuclear waste; montmorillonite; vermiculite; swelling capacity; erosion

1. Introduction

The establishment of a national repository (storage) for nuclear waste is one of the 2021 projects of the Malaysian Ministry of Science, Technology, and Innovation, which is projected to be completed by 2040 [1]. Nonetheless, the establishment of a repository with minimum hazard risk and economic implications is a significant environmental problem associated with this issue. Considerations for the development of a repository: characteristics of the waste to be deposited, engineering obstacles, geological context, and legal implications [2, 3]. As a result of the increase in radiation applications, particularly in the fields of medicine, industry, and technology, Malaysia's present intermediate waste storage is predicted to be fully occupied by 2025 [4].

Several countries, including China, Switzerland, Belgium, and France, have framework their radioactive waste storage systems by leveraging expansive clay or swelling clay such as Bentonite as natural sealing (or capping layer), buffer material, or engineered clay barriers in nuclear repository systems, particularly deep geological repositories [5, 6]. The applicability of these soils is determined by their unique qualities, which include very low hydraulic conductivity (impermeable), high swelling capacity, radioactive sorption capacity, microporous structure, and the ability to isolate waste for an adequate amount of time (≥100,000 years) [7].

Generally, expansive clay soil is composed of relatively fine particles (<0.002 mm) that are susceptible to volumetric changes (swelling-shrinking) that are directly related to changes in water content [8], with an abundance of expanding clay minerals such as montmorillonite and vermiculite [9]. The presence of swelling clay minerals in Malaysia is related to the type of host rock that undergoes an intensive weathering process. In other words, the distribution of expandable soil is widely available in specific natural environments in Sabah, highlighted by the weathering product of Mélange rock formation [10]. Consequently, this study aims to investigate the physicochemical and microstructural features of clay minerals and the possibility of these clays as natural sealing in nuclear waste respiratory systems, particularly for future near-to-intermediate type repositories (Figure 1).

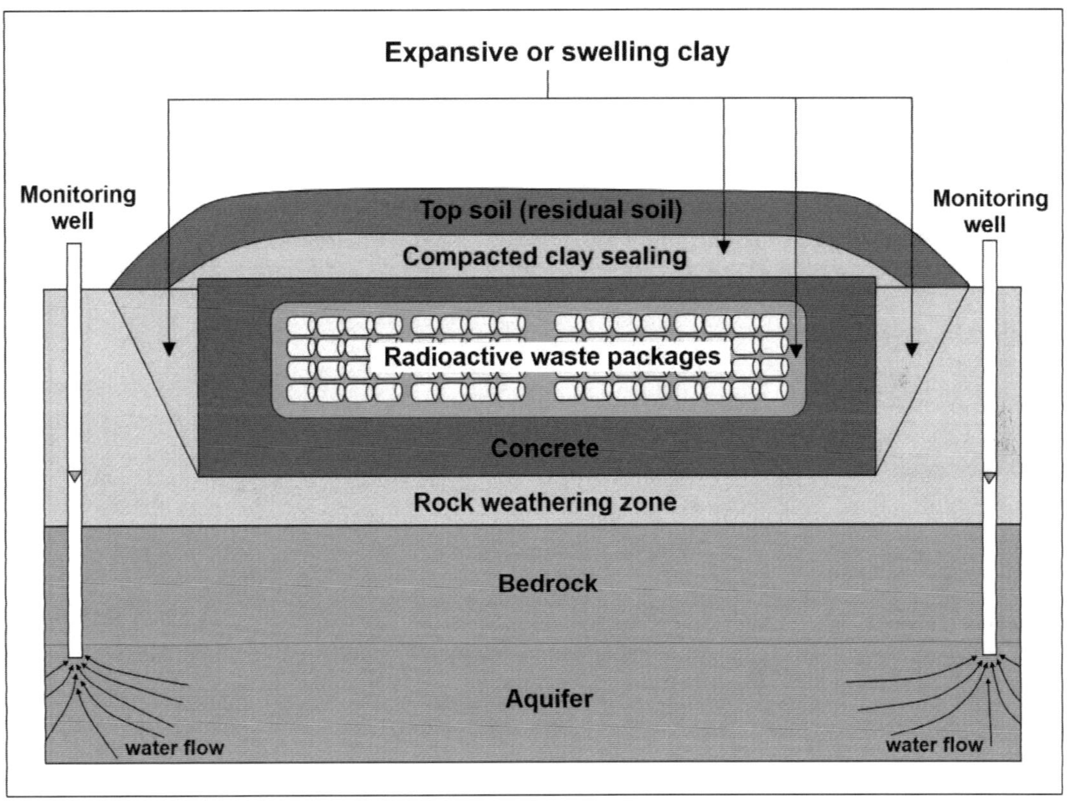

Figure 1. Schematic diagram of the near-surface repository.

2. Material and methodology

Two (2) samples of disturbed tropical clayey soil were collected for this study in Sabah, Malaysia. The physico-chemical characteristics of the soils were assessed using the following tests: soil specific gravity (American Society for Testing and Materials ASTM D854-14: Method A), pH test (ASTM D4972-01), particle size distribution analysis (ASTM D6913/D6913M-17/ASTM D7928-17), Atterberg limit/Consistency test (plastic limit, liquid limit test, and plasticity index) (ASTM D4318-17e1), linear shrinkage test (BS1377-1990, Test 5), and free swelling index (FSI) (IS 2720 (Part-40): 1977). By analyzing the microstructure morphology of clay minerals using Field Emission Scanning Electron Microscopy (FESEM) analysis and X-Ray Diffraction Analysis (XRD) with the DIFFRAC.EVA software, the type of clay mineral composition is identified. Based on the water flow rate (mL/s), effluent turbidity level, and diameter of the drilled specimen hole (mm), a pinhole dispersion analysis (a cylindrical shape with a diameter of 32 mm and a length of 38 mm) in accordance with ASTM D4647/4647M-13 can be used to assess the degree of dispersion of clayey soil according to the degree of dispersion, the analysis indicators range from dispersive (D1) (high erodibility) to non-dispersive (ND1) (resistance to erosion).

3. Results and discussion

Table 1 displays the index properties of samples of clayey soil. Compared to montmorillonite with a value of 2.60, vermiculite clay has a lower specific gravity value of 2.45, indicating that the soil is less dense and has a larger void for water to occupy. Each soil is classified as alkaline (pH > 7.0), which indicates poor texture, low microbial activity, and poor water infiltration capacity. In comparison to montmorillonite, vermiculite clay soil has higher LL, PL, and PI values and demonstrates higher soil consistency (LL>50%; PI>30%). However, montmorillonite clay exhibits greater clay activity than vermiculite, as determined by the ratio of PI and clay fraction, revealing the soil's swelling potential when reacting with water; the higher the value of A (Clay Activity), the greater the swelling potential. This value was supported by the FSI, which indicates that montmorillonite clay soil has a slightly higher percentage than vermiculite clay soil. According to the Unified Soil Classification System (USCS), montmorillonite and vermiculite clay soils were classified as sandy clay and clay, respectively, based on the particle size distribution (%) of sand, silt, and clay.

Table 1. Index properties of tropical expansive soil.

Index Properties		Montmorillonite Soil	Vermiculite Soil	Method
Specific Gravity, Gs		2.60	2.45	ASTM D854-14
pH Value		8.50	7.20	ASTM D4972-01
Particle size distribution	Sand particles (%)	70	56.8	ASTM D6913/
	Silt particles (%)	8	1.2	D6913M–17/
	Clay particles (%)	22	42	ASTM D7928–17
Plastic Limit, PL (%)		35.04	51.97	ASTM D4318-17e1
Liquid Limit, LL (%)		79.60	97.50	ASTM D4318-17e1
Plasticity Index, PI (%)		44.56	45.53	ASTM D4318-17e1
Clay Activity, A		2.03, Active	1.08, Normal	-
Swelling Capacity, FSI (%)		95	80	IS 2720 (Part-40): 1977

Linear Shrinkage (Ls) (%)	26	30	BS1377-1990
Soil Classification	Clayey Sand	Clay	USCS

Figure 2 depicts the results of XRD and FESEM analyses of soil. In accordance with the data, quartz (silica), montmorillonite, albite, and kaolinite predominate in montmorillonite clay soil. In contrast, vermiculite clay soil is dominated by vermiculite clay minerals and contains a negligible amount of quartz. At 30k magnification, FESEM analysis reveals that the microstructure of clay is predominantly composed of layered or stacked microaggregates formed by agglomeration of flaky morphology. Compared to montmorillonite, vermiculite clay has a notably flaky morphology and a porous structure. This morphology identifies the clay as phyllosilicate 2:1 type [11], which is typically distinguished by its flaky appearance [12].

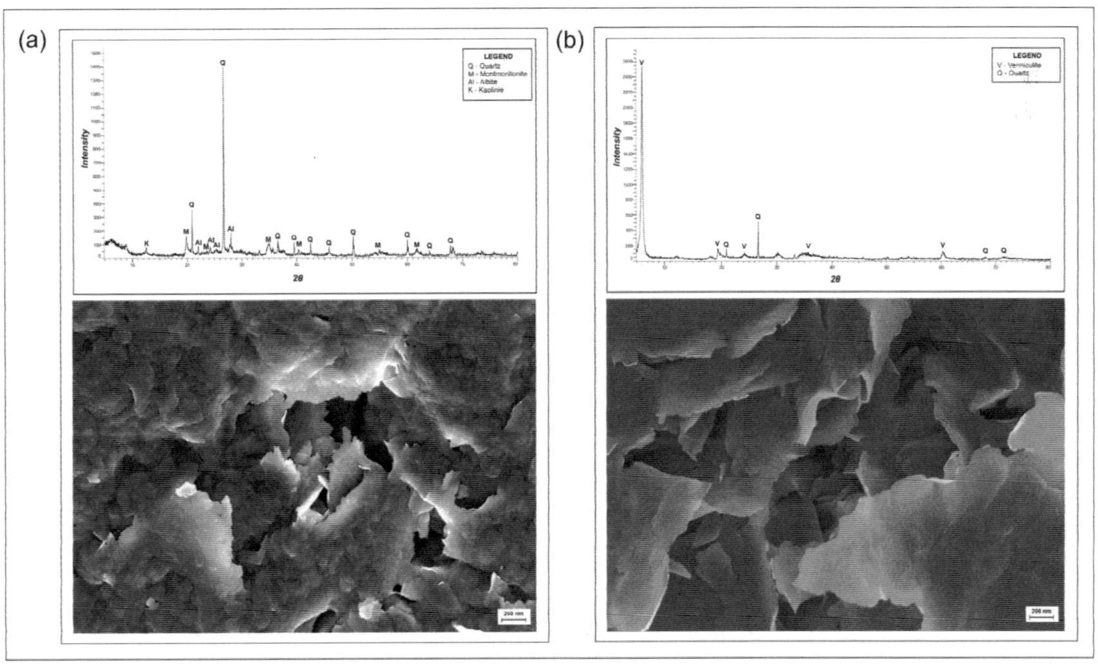

Figure 2. The figure illustrated XRD and FESEM analysis of soil samples: a) Montmorillonite soil sample; and b) Vermiculite soil sample.

Figure 3 represents the results of a pinhole dispersion test for analyzing soil dispersion. The size of the specimen hole is directly proportional to the effluent flow rate vs the duration of the test, with a greater erosion rate resulting from a higher effluent flow rate. The montmorillonite clay soil samples were classified as dispersive (high erodibility) (D1) with a final specimen hole size greater than 2 mm and a final effluent flow rate greater than 2.0 mL/s. Vermiculite clay soil is classified as non-dispersive (ND1) when the effluent flowing out of the specimen hole is completely clear and the change in diameter of the final specimen hole is 1.5 mm at a hydraulic height of 380 mm with a final effluent flow rate of 0.80 mL/s after 20 minutes. Due to its ability to disperse when reacting with water via reaction with monovalent cations (Na+), a Montmorillonite clay soil sample exhibited more extreme erodibility properties. The presence of higher Na+ ions is likely due to the presence of Albite mineral (Na-plagioclase-rich mineral) [13, 14].

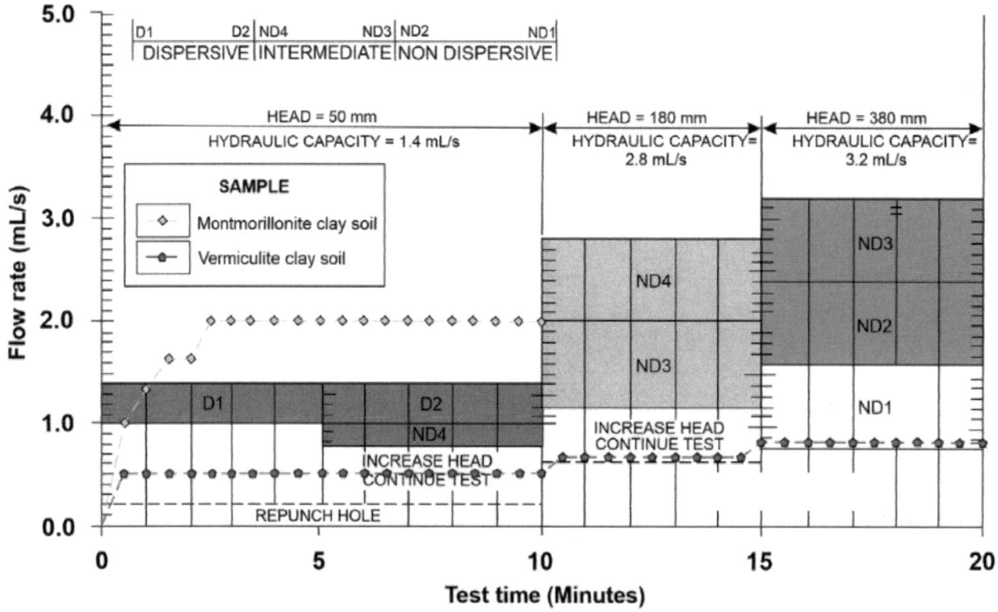

Pinhole dispersion analysis					
Soil sample	***Test duration (min)***	***Final hydraulic head (mm)***	***Effluent flow rate (mL/s)***	***Hole diameter (mm)***	***Soil dispersion level***
Montmorillonite	10	50	>2.00	>2	D1
Vermiculite	20	380	<0.80	<1.5	ND1

Figure 3. Pinhole dispersion analysis of soil samples as per ASTM D4647/4647M–13

4. Conclusion

In conclusion, expansive tropical soils dominated by montmorillonite and vermiculite have the potential to be utilized as a natural sealing in repository system due to their unique properties. However, montmorillonite clay tends to disperse when it reacts with water, resulting in severe erosion (tunnel/internal erosion), as opposed to vermiculite clay soil, which is resistant to erosion. Further research on the chemical (radionuclide adsorption capacity), geo-mechanical (maximum swelling pressure), and laboratory (self-healing ability from surface cracking due to desiccation) factors, emphasizing soil adaptability to tropical climates that are wet and humid annually is suggested to obtain more comprehensive data on the soil's suitability for a future national radioactive waste repository in Malaysia

5. References

[1] Ibrahim N, and Zakaria N 2021 Teknologi pelupusan repositori dekat permukaan: hala tuju baharu pelupusan sisa radioaktif di Malaysia *i-Nuklear* **2** 22–25.

[2] Norris S 2019 Multiple roles of clays in radioactive waste confinement – introduction Geological Society, London, Special Publications **482**(1) 1–9, doi:10.1144/SP482-2019-4.

[3] Wuqing Du W, You W, and Xu Z 2022 Review and prospect of legal development in commercial nuclear energy *Energies* **15** 4310 https://doi.org/10.3390/en15124310.

[4] Lim Y J, and Basri A 2022 Sites selection for a potential radioactive waste repository in Peninsular Malaysia: GIS-based weight linear combination and multi-criteria decision-making analysis *Prog. Nucl. Ener.* **149** 104252.

[5] Jalal F E, Xu, Y, Li, X, Jamhiri B and Iqbal M 2021. Fractal approach in expansive clay-based materials with special focus on compacted GMZ bentonite in nuclear waste disposal: a systematic review *Environ. Sci. Pollut. Res.* **28** 43287–43314 https://doi.org/10.1007/s11356-021-14707-7.

[6] Kale R C, Kapil B and Ravi K 2021 Response of compacted bentonite to hyperalkalinity and thermal history *Sci. Rep.* **11** 15483 https://doi.org/10.1038/s41598-021-95023-5.

[7] Liang S Y, Lin W S, Lin G F, Liu C W and Fan C 2021 The effect of porosity change in bentonite caused by decay heat on radionuclide transport through buffer material *Appl. Sci.* **11** 7933 https://doi.org/10.3390/app11177933.

[8] Jones L 2018 Expansive Soils. In: Bobrowsky P and Marker B (Eds.) *Encyclo. of Eng, Geol,. Encyclo. of Earth Sci. Ser.* (Cha: Springer) https://doi.org/10.1007/978-3-319-12127-7_118-1.

[9] Kumari N and Mohan C 2021 Basics of clay minerals and their characteristic properties. in: clay and clay minerals (pp. 1–29) IntechOpen Book Chapter, https://doi.org/10.5772/intechopen.97672.

[10] Raj J K 2021 Swelling Clay minerals and slope cut failures in the garinono formation along Jalan Sungai Hitam *Geol. Behavior* **4**(1) 29–34 http://doi.org/10.26480/gbr.01.2020.29.34.

[11] Islam A B M S, Khan Z H, Hussain M S and Uddin M M 2019 Scanning electron microscopic analysis of clays in the soils of lower Atrai Basin of Bangladesh *Dhaka Univ. J. Biol. Sci.* **31**(1) 105-115.

[12] Barton C D and Karathanasis A D 2002 Clay mineralsed Lal R *Encyclo. of Soil Sci.* 187-192 (New York, USA: Marcel Dekker).

[13] Tournassat C, Vinsot A, Gaucher E C and Altmann S 2015 Chemical conditions in clay-rocks. ed. Tournassat C *et al. Natural and Engineered Clay Barriers* 71–100.

[14] Shah A S N, Nazer N S M, Harris M I 2022 Morfologi Hakisan dan sifat serakan lempung kaolinit dan montmorilonit di kawasan tropika *Sains Malays.* **51**(12) 3879-3896, http://doi.org/10.17576/jsm-2022-5112-02.

AUTHOR INDEX

Abdullah, Intan Masayu 129
Abdullah, Nazifah 89
Adachi, Toshikazu 22
Ahmad, Muhd Izham 100
Ali, Noor Hasni M. 100
Ali, Nur Syazwani Mohd 70
Azman, Muhammad Azfar 114, 129
Bahrudin, Nurul Fairuz Dyana 129
Bakar, Asyraf Arif Abu 82
Chachuli, Fairuz Suzana Mohd 39
Chin, K F 7
Cioncolini, Andrea 1
Dixit, Tanuja 22
Diyana, Nurul Fairuz 114
Elias, Md Suhaimi 129
Ful, Chai Chan 70
Fuzi, Siti Aishah Ahmad 60
Hairaldin, Siti Zulaiha 100
Hak, Cik Rohaida Che 51
Hashim, Azian 114, 129
Hashim, I. H. 15
Hashim, Mohd Muzamil Mohd 114
Hashim, Zaredah 106
Hidzir, Norsyahidah Mohd 65
Hussain, Mohd Huzair 106
Iacovides, Hector 1
Idris, Faridah Mohamad 39
Idris, Mohd Idzat 96
Izzuddin, Izura 60
Jamaluddin, Khairulnadzmi 70
Jamaludin, Maziah 34
Karim, J A 7
Karim, Julia Abdul 106
Kawakubo, Tadamichi 22
Koh, Meng-Hock 15
Laili, Zalina 114
Mansur, Shahrina Akma 100
Maskin, Mazleha 106
Mejus, Lakam 114
Mohamed, Faizal Kunchi 106
Mohamed, Nurul Syazwina 89
Muhamad, Shalina Sheik 100
Mustaffa, Mohammad Syahiran 114
Nasir, Fairuz Mohd 89
Nasruddin, Nik Qurrotu Ain Ahmad 106
Nawawi, Nor Alia Aqilah 89
Noor, Norhazilan Md 82
Norazumin, Abdul Rahman 51
Omar, M R 7
Omar, Siti Aminah 114
Ong, J. F. 15
Paulus, Wilfred 114, 129
Pauzi, Anas Muhamad 1
Raduian, Nor Jawahir 129
Radzali, Nur Ain Mohd 65
Radzi, Afifah Mardhiah Mohamed 65
Radzi, Juliana Mohd 89
Rahim, Sapizah 60, 82
Razali, Azhani Mohd 34
Roslan, Nurliana 51
Sapiee, Nur Aqilah 129
Sarkawi, Muhammad Syahir 70, 82
Sazali, Muhammad Arif 70
Shah, A S N 148
Shah, Mohamad Shazwan Ahmad 82
Shukor, Shakirah Abd 114
Syazli, Farah Ainin Sofiya Muhd 96
Tajudin, Suffian Mohamad 89
Takayama, Ken 22
Tom, Phongsakorn Prak 106
Umar, Sarehati 82
Wah, Leo Kwee 22
Yaccup, Rahman 114
Yahya, Roslan 34
Zainudin, Nur Hamizah Mohd 89
Zali, M A 148
Zamani, Muhammad Nazrif 82